GROW YOUR OWN BOOZE

LORRAINE TURNBULL

Grow Your Own Booze

Copyright

Copyright© 2022 by Lorraine Turnbull

Fat Sheep Press, 46300 Milhac, France.

ISBN 978-1-7396072-2-7

For my son.

Even you can make some of the easiest
recipes in this book!

Acknowledgements

Writing a book like this comes after many years of creating and drinking some of the best and worst of alcoholic beverages. I especially thank Tom Oliver for introducing me to the best perry I've ever drank at his fantastic farm at Ocle Pychard in Herefordshire; and John Cooper who bullied me into making cider commercially many, many years ago, for which I will always be grateful.

I'd also like to thank everyone who has contributed recipes or undertaken some experimental brewing for me, and my ever-fantastic reader and friend Rod Haselden.

About the author

Lorraine Turnbull wanted to be a farmer since she was five years old. In her mid-forties, she uprooted herself and her family and moved to a run-down bungalow with an acre of land and an Agricultural Occupancy Condition in Cornwall. She retrained as a teacher, worked as a Skills Co-ordinator for The Rural Business School, and started commercial cider making in 2010.

In 2014 she won the Cornwall Sustainability Awards *Best Individual* category, and after more than a decade of living the 'Good Life' on a Cornish smallholding, she made a huge leap of faith and moved to France in 2017. She now lives on a small farm with an orchard and some sheep with her husband and Cocker Spaniels. Her first book was published in May 2019.

Lorraine has been making beer, country wines and liqueurs for home consumption since her twenties and has been making cider since 2008. She made the decision to begin commercial cider production in 2011, and made a range of still and bottle conditioned full-juice ciders under the brand Spotty Dog Cider.

Other books by the author

The Sustainable Smallholders Handbook (2019)

Sustainable Smallholding (2020)

Living off the Land: My Cornish Smallholding Dream (2020)

Mum's the Word (2021)

How to Live the Good Life in France (2nd edition 2022)

Murder at the Moulin (2022)

Connect with Lorraine:

Facebook   https://www.facebook.com/LorraineTurnbullAuthor/

Twitter   @LorraineAuthor

Instagram   lorraineauthor

Youtube: Lorraine Turnbull Smallholder & Author

Contents

INTRODUCTION

I've made my own beer, wine, cider and liqueurs as part of my sustainable smallholding lifestyle since 2005 (although I have actually made quite a lot since my twenties). Part of the reason has been financial: smallholding can be a delightful way of life, but there is nothing more rewarding than making your own enjoyable and delicious alcoholic drinks and knowing it saves money. Another reason has been to avoid the many additives and chemicals that seem to abound these days in commercially made foods and drinks. There are rising numbers of the population allergic to sulphites, gluten, and concerned about the high sugar content of flavoured high alcohol beverages, so popular amongst the younger population. I have to add that there is something satisfying about taking some easily home-grown or foraged ingredients and producing a really delicious tipple that you have made with your own hands.

Many readers may be aware that I was in fact a commercial cider producer for many years in Cornwall. Spotty Dog Cider was a huge success, both commercially and sustainably, and since closing the business and moving to France in 2019 to retire, I have planted a new orchard and continue to produce juice and cider, amongst other drinks for my own consumption here.

Now, with a cost-of-living crisis in the UK, and unrest on the world stage threatening food shortages and high prices, I've decided it was time this little book was finally published in order to enthuse newbies in the home-brew game. Home production of alcohol isn't new, but existing books on the subject have been a little old-fashioned and concentrate on the production of the drink itself. This book aims to link producing alcohol and growing or foraging for the raw ingredients, so even a city dweller can still go out, gather ingredients and start those first steps towards self-sufficiency.

I've divided the book into two sections, which you can dip into to learn how to grow suitable plants, and how to use these in your alcohol production. I have not included home distillation of spirits in this book – it's outwith my personal experience and is sadly illegal in both France and the UK, unless you are officially registered (in the UK with HMRC).

I encourage the reader to give at least one of the easy recipes a try; there really is nothing like making your own tasty tipple. Keep a notebook detailing amounts, dates and the method you've used and if it doesn't turn out perfect the first time, give it another go - practice makes perfect. Humans have been making alcohol for thousands of years. This book aims to help the beginner understand how to make decent alcohol, and explains some of the technical and scientific terms that are off-putting.

Finally, remember alcohol is a drug and highly addictive, so please do drink responsibly.

PART 1

EQUIPMENT

The equipment you will require to begin making your own wine, cider, beer or liqueur will be decided by what you choose to make; but many items are standard to all. Some things, like a hydrometer (used to work out the final alcohol strength of whatever you make) or airlocks can be purchased from a home-brew shop or increasingly on the internet. Demijohns (large glass or plastic brewing containers) can be found at car-boot sales, or again from specialist home-brew suppliers. Accurate kitchen scales are always necessary, as are potato mashers, large spoons, a measuring jug and funnel. When we get to making cider then some of the equipment starts to increase dramatically in size and cost. You will need some sort of scratter or pulper, a press of some sort, and large fermentation vessels.

I've subdivided this section into general sections for beer, wine and cider; and as you read the book, you will see that experience and or necessity are the mother of invention, and things that can be used for one product can be used or adapted for others.

Products such as liqueurs will simply require surplus clean glass screw-top bottles; so if you wish to make sloe gin, schnapps, or flavoured vodka, simply re-use clean spirits bottles, sterilising thoroughly inside the bottle and lid before use.

Equipment for beer and lager

Think about where you are going to make and store your final product. To make a half-decent amount you'll need bottles for the finished product or a pressure barrel. A cool storage area is key (or you can do what we do here in the hot South West of France- and keep beer making confined to the cooler months of late autumn or spring time). I've listed equipment here to make around 40 pints; the standard beer kit quantity, and suggest you start with a readily available beer kit, which you can buy from your home-brew shop or from the internet.

Ready to make beer/lager kit (this usually comes in a tin).

Food Grade plastic bucket with lid of 30 litre capacity, with bung & airlock.

If you want to do quick beer making then a **BIAB straining/mashing bag** is a good investment.

Large plastic spoon or paddle.

Muslin cloth, straining bag or coffee filter papers.

Hydrometer & hydrometer jar or tube - for any home-brew enthusiast this is a must-buy. Get an easy to read long hydrometer and store safely in a padded tube out of harms way. They don't cost much and are invaluable to correctly tell you the starting sugar level and the end fermentation level and you can easily work out the ABV of your product (whether it be wine, beer or cider).

Campden tablets or tablets for sterilising baby feeding equipment are widely and cheaply available. For those with sulphite allergies, go for the latter, which you can buy in the baby section from your local supermarket.

Syphons - You can use a simple plastic tube siphon from a high surface to a lower or use a pump action siphon. A reasonable length of tubing (at least 1.5 metres) is preferable to a shorter tube. Again, food grade is best and flushing through with clean water and a sterilising fluid essential.

Measuring jugs and funnels - Food grade plastic are good, but the measurements must be clearly and accurately marked. There is nothing worse than having to squint to read the levels, especially if you are not in a well-lit area or dealing with dark liquids or spillages. Ensure everything is scrupulously clean before use.

Glass bottles and crown caps (brown bottles are traditionally used for beer or ale, and green bottles for lagers, but of course you can use clear bottles as they are frequently cheaper). Glass bottles can be recycled again and again. Crown caps need a capper tool to crimp them down tightly onto a bottle and cannot really be re-used.

Crown capper tools - When I began making cider I invested in a crown capper, realising that I could not only use it for my cider but also for our home made beer and lager. Widely available on the internet, or if you have a fellow home-brew enthusiast nearby, you can share. They are easy to use and clean.

Bottle drainer tree - usually made from plastic and suitable for all glass bottles. The bottles are hung to drain upside down on plastic pegs set in decreasing circles as they rise. These can be purchased fairly cheaply from most homebrew suppliers or internet sites.

Storage crates - I use large, shallow plastic crates for all my home brew and cider. I label the crates themselves to differentiate between products (and thus can reuse glass bottles without having to remove sticky bottles labels). They stack well, and can be hosed clean. You can use cardboard crates, but in my experience they get damp unless stored well, and lose their strength over time.

Equipment for wine making

Food Grade plastic bucket with lid of 25 litre capacity.

A large saucepan or preserving pan.

Fine straining bag

Hydrometer & hydrometer jar or tube - for any home-brew enthusiast this is a must-buy. Get an easy to read long hydrometer and store safely in a padded tube out of harms way. They don't cost much and are invaluable to correctly tell you the starting sugar level and the end fermentation level and you can easily work out the ABV of your product (whether it be wine, beer or cider).

Campden tablets or tablets for sterilising baby feeding equipment are widely and cheaply available. For those with sulphite allergies go for the latter, which you can buy in the baby section from your local supermarket.

Syphons - You can use a simple plastic tube siphon from a high surface to a lower or use a pump action siphon. A reasonable length of tubing (at least 1.5 metres) is preferable to a shorter tube. Again, food grade is best and flushing through with clean water and a sterilising fluid essential.

Sachets of wine yeast - There is a huge variety available (5 g sachets do 23 litres or 5 gallons).

Measuring jugs and funnels - Food grade plastic are good, but the measurements must be clearly and accurately marked. There is nothing worse than having to squint to read the levels, especially if you are not in a well-lit area or dealing with dark liquids or spillages. Ensure everything is scrupulously clean before use.

At least three **1 gallon demijohns** or 3 25 litre food grade fermenters

Airlocks and bungs to fit.

Glass bottles and screw top lid or corks (clear bottles for white wines and green for red is traditional). Glass bottles can be recycled again and again. Screw top lids can be reused, corks and crown caps cannot. If you intend to make sparkling wines you need to consider champagne bottles, corks and cages, and store in a very cool place.

Corking machine - To make most wines you will need to consider either traditional type corks or screw tops. If you choose traditional corks then you will need a cork inserting machine. Available on the internet, or if you have a fellow home-brew enthusiast nearby, you can share.

Bottle drainer tree - Usually made from plastic and suitable for all glass bottles. The bottles are hung to drain upside down on plastic pegs in decreasing circles as they rise. These can be purchased fairly cheaply from most homebrew suppliers or internet sites.

Storage crates - I use large, shallow plastic crates for all my home brew and cider. I label the crates themselves to differentiate between products (and thus can reuse glass bottles without having to remove sticky bottles labels). They stack well, and can be hosed clean. You can use cardboard crates, but in my experience they get damp unless stored well, and lose their strength over time.

Equipment for cider making

Obviously this will depend on the scale of the cider making, but as this is a book aimed at beginners, I will list accordingly.

Washing up bowl - a large washing up bowl, or baby bath will suffice to wash apples.

Press - Please see the appendix for plans for a small wooden rack and cloth press. You must dismantle and thoroughly clean this as soon as you have finished using it as the acid in the juice will begin to eat into the metal. A larger press may be hired (or bought) or you could visit a pressing service.

Scratter - For small quantities a home food processor will quickly pulp your apples; again wash thoroughly afterwards.

Food Grade plastic bucket with lid of 25 litre capacity.

Hydrometer & hydrometer jar or tube - for any home-brew enthusiast this is a must-buy. Get an easy to read long hydrometer and store safely in a padded tube out of harms way. They don't cost much and are invaluable to correctly tell you the starting sugar level and the end fermentation level and you can easily work out the ABV of your product (whether it be wine, beer or cider).

At least three **1 gallon demijohns** or 3 x 25 litre food grade fermenters.

Airlocks and bungs to fit.

Campden tablets or tablets for sterilising baby feeding equipment are widely and cheaply available. For those with sulphite allergies, go for the latter, which you can buy in the baby section from your local supermarket

pH strips (optional).

Sachets cultured cider yeast (5 g sachets do 23 litres or 5 gallons).

Sachets yeast nutrient (between 50-100 g).

Syphons- You can use a simple plastic tube siphon from a high surface to a lower or use a pump action siphon. A reasonable length of tubing (at least 1.5 metres) is preferable to a shorter tube. Again, food grade is best and flushing through with clean water and a sterilising fluid essential.

Measuring jugs and funnels - Food grade plastic are good, but the measurements must be clearly and accurately marked. There is nothing worse than having to squint to read the levels, especially if you are not in a well-lit area or dealing with dark liquids or spillages. Ensure everything is scrupulously clean before use.

Bag in Box storage - available in 3, 5, 10 or 20 litre capacity these food grade pouches come with a tap to open and sit inside cardboard packaging. Once filled the bags can be pasteurised at 70 °C for 20 minutes and offer an alternative to glass bottles.

Glass bottles and crown caps (clear is the normal colour for cider). Glass bottles can be recycled again and again. Crown caps need a capper tool to crimp them down tightly onto a bottle and cannot really be re-used. They come in a range of colours allowing you to differentiate batches by colour – e.g. black cap for still, red cap for sparkling etc.

Crown capper tools - When I began making cider I invested in a crown capper, realising that I could not only use it for my cider but also for our home made beer and lager. Widely available on the internet, or if you have a fellow home-brew enthusiast nearby, you can share. They are easy to use and clean.

Bottle drainer tree - usually made from plastic and suitable for all glass bottles. The bottles are hung to drain upside down on plastic pegs in decreasing circles as they rise, and these can be purchased fairly cheaply from most homebrew suppliers or internet sites.

Storage crates - I use large plastic crates for all my home brew and cider. I label the crates themselves to differentiate between products (and thus can reuse glass bottles without having to remove sticky bottles labels). They stack well, and can be hosed clean. You can use cardboard crates, but in my experience they get damp unless stored well, and lose their strength over time.

How to use a hydrometer

A hydrometer is a basic tool that measures the amount of sugar in a liquid, and is probably the most essential piece of equipment to make beer, cider and wine successfully. They are not expensive for the quality of the results they give. The more sugar there is the higher the reading will be (at the start of fermentation), and the more alcohol there is in the liquid the lower the reading will be (at the end of fermentation). It's used to measure the starting S.G. (Specific Gravity) of a liquid, measure against the finishing S.G. and calculating the ABV (Alcohol by Volume) of the finished product.

It should be used whenever you are making any alcoholic drink – wine, beer or cider. It's not necessary to use when flavouring or infusing spirits such as gin, vodka or rum.

Take an initial S.G. reading when you first press your juice or put your beer making liquids together, and note it in your notebook. All hydrometer readings are affected by temperature, so try do take the Initial, Final and any readings in between at the same room temperature and liquid temperature.

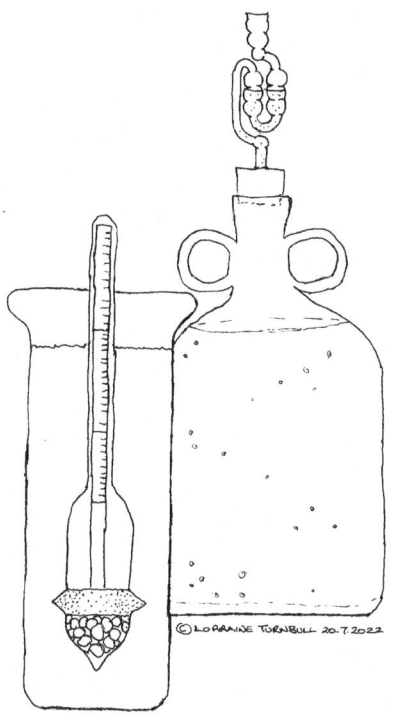

How to take a reading

I like to use my hydrometer with a hydrometer or sample jar. I keep them very clean and stored carefully as they are easy to knock over. I also put my hydrometer jar on a flat surface at eye level with a white background behind to better read it.

A normal hydrometer jar is 200 mm tall. When taking a sample fill it to about 35 mm from the top and *gently* drop the hydrometer into the sample liquid. Give the hydrometer a gentle spin with our finger and thumb – this loosens any air bubbles clinging to the hydrometer, which could give a false reading.

Look to where the hydrometer floats on the liquid. To get an accurate reading read the level at the bottom of the meniscus – the level of liquid in the centre of the liquid level (the liquid will be pulled by surface tension to the sides of the sample jar and the hydrometer itself).

For a typical beer your starting Specific Gravity (your initial S.G.) should be (and there are so many variations that I can only suggest you use this figure as a guide) between 1.045 and 1.060. As your ferment progresses and turns from 'wort' to beer, you may wish to take further S.G. readings to check that the sugar is being turned to alcohol, and the gravity reading is falling. A Final Specific Gravity reading for beer could be between 1.015 and 1.005 (typical pub 'session' beers are around 4.5 % ABV).

When making wine your Initial S.G. reading should be between 1.070 and 1.090. The Final S.G. reading for a dryish wine will be about 0.990, and 1.005 for a sweet wine. As a guide an Initial S.G. of 1.070 will finish with an ABV of 10.5 %, and with an Initial S.G of 1.090 you could expect an ABV of 13 %. Wines should be fermented to dry and then sweetened after stabilising (treating the finished fermentation with Campden tablets). You can then sweeten with grape juice or sugar.

Calculating ABV

It's not essential in many cases to know what your final ABV may be, but some recipes will suggest taking readings for successful fermentation. Subtract the Initial S.G. reading from the Final S.G. reading and divide by 7.362. So – imagine you have a wine which the Initial S.G. reading is 1.080, and finishes at 0.990, then the difference is 90 points. Divide this by 7.362 and the ABV of that wine is 12.23 % ABV.

How to fit an airlock

An airlock is a handy device used by many wine, cider and beer makers to allow gases to be released from the fermentation whilst preventing the contamination by oxygen, and insects, wild yeasts and dirt. It is in its simplest form a one-way valve.

Remember that the fermentation vessel is sealed and as the yeast ferments the sugars in the liquid they release gas as a by-product. This gas rises through the liquid and as the pressure in the vessel expands is forced as a bubble through the airlock.

Ensure your airlock is clean and sterile as is the bung that you will insert into your fermentation vessel.

Half-fill the airlock with water, force the airlock into the bung and then fit the bung snugly into your demijohn or fermentation vessel.

When fitting an airlock at the start of fermentation it is advised that you leave reasonable headroom between the top of the liquid and the bottom of the airlock. This will prevent a violent fermentation filling the airlock with foaming liquid. In a demijohn this is why a lot of recipes suggest filling to the shoulders rather than filling to the neck. When you rack off into a clean vessel, always replace the airlock with a clean sterilised one. Now you can top up the alcohol in this new vessel to the neck (with either water or some finished alcohol) before fitting the new airlock. Check occasionally that the water has not evaporated.

Racking off

Racking off is the process of taking your finished beer, wine or cider from the spent yeast and sediment (the lees) and housing it in a clean vessel. The racked-off alcohol can then mature.

It may be necessary to rack-off more than once if your beer, cider or wine throws a lot of sediment.

When I am ready to rack-off, I like to carefully and gently move my fermentation vessel the night before to a cold place if possible, with the full demijohn on a shelf or table and the clean vessel on a lower surface or on the floor. With demijohns this is easy, but the larger the vessel, the more problematic. I've had cider fermenting in 1000 litre IBC's in my cider shed, and as I could not realistically move them, I simply threw open the doors the evening before and waited for the temperature to drop (I always racked off my cider in late autumn or early spring). The cold is useful to force the yeast and sediment down to the bottom of the fermentation vessel and helps clear the alcohol before racking.

Remove the airlock and bung carefully, and insert the syphon tube, and draw off about 80-90 % of your finished drink into a sterilised vessel. Take care to keep this end of the syphon fully submerged in the liquid, but above the lees. Do NOT be greedy; and leave enough liquid on top of the sediment (the lees) to ensure you don't syphon some off by mistake. Also you must exercise care when siphoning to place the other end of the syphon tube into the bottom of the new container to slowly run in the racked-off liquor whilst minimising air contact as much as possible.

Of course, you'll realise immediately that you now have less liquid, and if you are using demijohns you must top this level up carefully by adding some water to the neck, and fitting another airlock for a few days to check that the process hasn't encouraged a secondary fermentation as you have disturbed some yeast.

Why do we have to rack off? Well, it helps to clear your wine, cider or beer, but also prevents the spent lees from adding off-flavours. If you are unhappy with the clarity of your racked-off wine you could add some finings or pectolase; personally, I'm unhappy to add finings as I like my alcohol as natural as possible. Every time you rack, you risk exposing your product to oxygen. In my experience less is more and this is why I rack early morning on a cold day. Plan your racking so that you are not harassed and can take time to rack off slowly.

Also, you will lose some of your finished alcohol when you rack off. If you are making in bulk you could keep a vessel solely for topping up your main fermenters with already fermented surplus; I used to do this with cider. Or you can add water. Or, if you stabilise your finished product with crushed Campden tablets then you can add fruit juice or sugar water. Remember racking does NOT stop fermentation, it merely slows it down.

MATURATION

Yes – this is in huge letters because it's important. *Of course*, you'll be eager to try your beer, wine or cider or whatever as soon as possible, but unless the recipe says it's ready to drink in a week, PLEASE let it sit and mature. It really will be much nicer to drink. This is the main reason why people give up on making their own booze.

Cider is generally ready to drink after 3 months maturation, but BETTER after 6 months.

As a general rule for wine making, allow one year for flower wines, two years for fruit wines and three years for vegetable or grain wine.

If you are infusing spirits, such as schnapps, sloe gin, etc allow normally at least a week to mature after infusing, but whatever you make will taste better as time goes on, so perhaps make enough for two batches; one for drinking now, and one for later.

There are countless recipes and variations of country wines, quantities can be varied and the strength and sweetness of any recipe can be adjusted. Most of the recipes in this book are in gallons for wine, because a demijohn holds a gallon. If you want a richer flavour or colour, simply increase the amount of the main ingredient; if you wish the wine to be sweeter increase the sugar, if less sweet reduce the sugar. Remember that when making fruit-based wines a lot of the sugar is in the fruit itself, and that when you're making flower and vegetable based wines you'll need to add more sugar. And it's a lot easier to sweeten a finished wine (with sugar or a non fermentable sweetener) than to make an over-sweet wine drier.

FORAGING

Our ancestors have foraged for food, berries and other ingredients since the dawn of time, and today it is again becoming popular as people like to forage seasonally and locally either through enjoyment or financial necessity; in fact some enterprising individuals have made good small businesses supplying elite restaurants with foraged wild herbs, mushrooms, edible seaweeds and truffles.

You will need to ensure you are clear on a few matters before you start to do this yourself. Firstly, get the permission of the landowner. You CANNOT just take fruit, berries or flowers or dig up plants from someone else's property. Of course, in the countryside or remote areas no-one is going to mind you collecting a bucket of brambles or some sloes from a hedgerow, but be aware of your legal responsibilities. There is also a list of protected wildflower species that are strongly protected by law and prohibited from collecting under any circumstances. You will not find any recipes for them in this book. In the UK the law; *The Wildlife and Countryside Act 1981*, Schedule 8 says that it is illegal to dig up or remove a plant (including algae, lichens and fungi) from the land on which it is growing without permission from the landowner or occupier. Some species are especially protected from picking, uprooting, damage and sale. The list can be found here: **https://www.legislation.gov. uk/ukpga/1981/69/schedule/8**

The Foragers Calendar

This calendar contains some common wild plants that grow in a temperate climate (most of the UK, northern to central France, USDA zone 8/9). Remember with climate change any certainty is out of the window.

	JAN	FEB	MAR	APR	MAY	JUN	JUL	AUG	SEP	OCT	NOV
Nettles	█	█	█	█	█	█	█	█			
Dandelion	█	█	█	█	█	█	█	█	█	█	
Gorse				█	█	█	█	█	█		
Broom				█	█	█					
Elderflower					█	█					
Honeysuckle					█	█	█				
Oak leaves				█	█						
Beech leaves				█	█						
Wild strawberry					█	█	█	█	█		
Roses				█	█	█	█	█	█		

30

Green walnuts	
Meadowsweet	
Brambles	
Heather	
Elderberries	
Rowan berries	
Wild raspberries	
Hops	
Burdock	
Crab apples	
Hawthorn berries	
Rosehips	
Sloes	
Damsons	

You also need to be absolutely sure you can confidently identify the species you are going to forage. For example, elder (*Sambucus nigra*) is perfectly safe and non-toxic, but the very similar *Sambucus ebulus*, the Dwarf Elder or Danewort plant is poisonous. Seemingly innocuous wildflowers can also lure you into a false sense of security –many fungi, and some flowers etc are poisonous. Stick to the recipes in this book or do a lot of internet research.

Foraging is also seasonal, so you will only be able to collect elder flowers, for example in May or early June and elder berries in late summer, but remember you can freeze them (label everything). I have included a foraging calendar in the appendix, but obviously this will depend on your geographical location, and the seasonal abnormalities so frequent due to climate change.

When you decide to go foraging, prepare for the weather, take some suitable containers, take a knowledgeable and clear reference ID book (especially if you are searching for fungi). In France, where collecting wild mushrooms is pretty common, you can take your fungi to any local pharmacy where they will safely identify your fungi; if you remember to keep different specimens separate. Remember to avoid foraging at busy roadsides, where chemical emissions from vehicles and fuel and oil residues may contaminate your plunder; and anywhere where dogs may have fouled.

GROW YOUR OWN BOOZE PRODUCING PLANTS

When I started thinking about producing this book, it began as a simple book on how to grow apple trees and produce cider. Of course, the apple is a really versatile fruit, so I had to add information about juicing, and verjuice and then it seemed churlish not to include perry, and then other orchard fruits. Very soon, the project had taken on a life of its own; and as I really wanted to explain both how to *produce* these drinks and how to *grow* the plants themselves, the book has expanded somewhat.

This section, although extensive in the range of plants covered , does not claim to be exhaustive, and simply intends to highlight the wide variety of plants that can be reasonably grown in the UK and temperate areas of the world and used to produce your own alcoholic (and non alcoholic) drinks. I have not gone into minute detail of 'how to grow', because there are more than one way to grow plants, and any good gardening reference book or a trawl on the internet will provide reams of information.

In my own experience I have to say that I began gardening as a complete novice at the age of nine, and now at the age of 60, can say that although I have loads of experience, I'm still learning something new every day. I've grown plants on a windowsill in an apartment, in a porch, in growing-bags, in a garden and then developed a couple of smallholdings. Now in my golden years, I've reduced my vegetable gardening to raised beds, but still have a low-maintenance orchard.

I'm acutely aware that we don't all have access to the same amount of land or space, and that we are restricted by location and climate as to what we can reasonably grow. Finally, there are so many different things out there to grow that you can turn into alcohol that sometimes it's overwhelming to know just where to start.

I'd start by answering a few simple but important questions –

- How much space and how much growing experience do you have?

- What can I grow that will save me money?

- What will give me the best results and the most pleasure?

There is no point having grandiose plans if you only have space for a few pots on a balcony or trying to grow exotic and difficult species if you've never grown anything before or live in a challenging climatic location. Start small and easy and expand as you gain confidence. If you're hampered by a poor climate then perhaps you can share a greenhouse or even have pots or growing bags in a sunny porch.

Also, some ingredients can be bought so cheaply and are widely available in shops or supermarkets or can be foraged for free that there is no point in growing them – concentrate on expensive or hard to source items. I'm never ever going to grow my own potatoes when they are 'cheap as chips' to buy from a supermarket.

Finally, the *real* fun in making your own booze is the end result – tasty booze. If you drink wine, then make your own wine; if you are a beer fan, try some of the huge varieties of beers. Grow the things that will please you, and your taste buds. Beer makers might grow their own hops, and try adding some home-grown fruit to their beers; cider makers might experiment with adding some raspberry juice to their cider; and wine makers – oh my goodness, you can grow rhubarb to add body to any floral wines to jazz them up, or grow some ginger in a pot to add to a winter elderberry wine, or even grow some chillies and try a small amount of this in some of the more robust vegetable wines, such as parsnip or carrot. You're only going to be limited by your imagination and your unwillingness to try something new. I'm going to start this section on plants with one of the easiest plants to grow.

APPLES

Apples are one of the easiest fruits to cultivate and grow. With a choice of rootstocks to restrict growth and many varieties to choose from, you could literally pick apples from late August through to the end of December, filling your freezer with fruit and your store cupboards with juice and cider, cider vinegar and verjuice. Even the smallest balcony can be home to a small apple tree in a large pot, and larger gardens can have their very own orchard. For those who do not wish to grow their own apples, you can buy some from your supermarket, farm shop, direct from a local producer or beg from a friend with surplus apples. When I first started Spotty Dog Cider I offered a collection service to all the small orchards nearby who were just leaving apples to rot on the orchard floor! Cider and apple wines can be made from *any* type of apples, but you will get a more complex range of flavours if you mix varieties and types. Go easy on including cooking apples when making wine or cider as they are pretty sour.

As this book is primarily concentrating on using fruit and vegetables for making alcohol, your choice of apples will need to be examined. If you only have space for one tree, whether it's in the ground or in a large container, this needn't stop you having a family tree with two or even three varieties grafted onto it. You'll pay more (unless you graft this yourself), but it will pollinate better and give you a choice of fruits if you look after it. So in this case you might choose one dessert variety and one dual purpose, or a dual purpose and a cider variety; whatever floats your particular boat.

If you have space for a few dwarfing or semi-dwarfing trees or espaliers then again, choose carefully. The table below shows a range of rootstocks and their attributes.

APPLES				
Rootstock	M27	M9	MM106	M25
Ultimate height	1-2m	2-3m	3.5-5m	6.5-9m
Spacing	1.5 - 2m	1.5 – 2m	3.5m	7.5 – 9m
Years to fruiting	2-3	2-3	3-4	6-7
Uses	Cordon, pots (patio or balcony)	Cordon, espalier, half-standard (small garden)	Espalier, garden standard (average garden)	Farm (farm/estate/park)
Yield at 10 years	7kg	20kg	50kg	120kg
Comments	Permanent stake Good soil No grass	Permanent stake Good soil No grass	Stake for 5 years Ordinary soil Can be grown in grass	Stake for 5 years Ordinary soil Grown in grass

So you can see the rootstock makes a big difference to the size and vigour of the tree. The rootstock is the root-mass and the stem above for around 16 - 20 cm, then the main part of the stem and the crown, the flowers and fruit is the grafted variety. You'll need to choose the 'form' of the tree. A cordon is basically a straight trunk from which flowers and fruit should appear from small spurs. An espalier is for training either against a wall or as a step-over. There are usually two main branches which have been trained horizontally. A half-standard and standard are the typical 'lollipop' shaped trees. A half-standard has the branches coming from the trunk around 1.5-1.8 m high, and a standard has branches starting at around 2 metres.

Then you have to choose your varieties – well, I could write a whole book on this alone, so will refer you to the internet with some advice. Choose disease resistant varieties where possible; choose varieties that bloom at the same time (so they pollinate each other) and avoid varieties with scab or canker. Now this is controversial, because *everybody and their granny* want to have a *'Bramley'*; and they are notorious for canker, especially in damper Western and Northern areas. If you had space for three trees only, I'd choose an early eater (like *'Discovery'* or *'Katy'*), a late eater or dual purpose variety (like *'Reinette d'Orleans'*) and a bittersweet cider apple (see the section below). They can ALL be used for juice, cider and apple wine.

If you fancy growing some cider varieties you can buy or graft your own, and I've listed some popular varieties below. Remember you don't have to *have* these in your mix, but they will add a more complex flavour, to both juice and cider. If you could only choose one cider tree, personally, I'd go for a bittersweet, and a variety that didn't have problems with canker or scab (more of an issue in damp, humid conditions). There are many, many varieties of apple; trust me on this; I'm a bit of an apple anorak, so you'll always find one to suit your location and conditions.

Suitable cider apple varieties

Browns apple	Cider	Early	Sharp
Tom Putt	Cider	Early	Sharp
Dabinett	Cider	Late	Bittersweet
Yarlington Mill	Cider	Mid/late	Bittersweet
Slack ma Girdle	Cider	Late	Sweet
Harry Masters Jersey	Cider	Mid/late	Bittersweet
Hangy Down	Cider	Early/mid	Bittersweet
Foxwhelp	Cider	Early/mid	Bittersharp

Pruning takes place twice a year; in winter for cutting out dead, diseased and damaged parts and to reduce congestion in the centre of the tree. Aim for an open goblet shape, and go easy – there is always next year. Summer pruning gets rid of excess growth and non productive water-shoots. Do not paint anything onto the cut surfaces, and use sharp secateurs or loppers that you've wiped with methylated spirits or white spirit prior to going to each tree.

When you are planning and planting your trees try to avoid steep slopes (for safety) and frost pockets (can reduce the harvest considerably), and plant and back-fill with the soil that came out of the hole. You shouldn't add compost, but can sprinkle some general fertilizer if your soil is poor. Stake the tree and protect from rabbits and deer if you have them. Water generously after planting and at any time in the first year if you think the tree needs it. A good long soak every few days is better than a dribble of water daily.

When your trees have reached their fourth or fifth year you may notice 'June drop', which is the tree shedding some excess fruitlets. You can mimic this behaviour by reducing fruits to two per spur; rubbing out the 'king fruit' in the centre of the truss. Depending on the variety your fruit will be ripe from August through till December. You can test the ripeness by choosing one apple, seeing if it comes away lightly in your hand when you cup it, and further checking by cutting it open to reveal the pips. Brown pips are ripe, white pips are not. Strong winds may shake down some windfalls. If they are not ripe you can store them for a week or two in shallow trays with a little straw between them or individually wrapped in newspaper, out of the reach of vermin.

If you'd like to try growing your own apple tree rootstocks and then grafting your own varieties onto them, then here is the way to do it. I also have some videos about this on my Youtube channel **Lorraine Turnbull Smallholder & Author.**

You'll need a weed-free prepared nursery bed or some large pots filled with soil-based compost. Choose what rootstock you want to propagate. I use MM106; these produce semi-dwarfing trees and have some disease resistance too. You'll have to initially invest in a few rootstocks bought from a reputable supplier to start your own supply; try to buy the longest you can get – you'll need at least 15 cm long. Unpack them as soon as they arrive, and soak in a bucket of water for an hour or so, before planting or potting up.

You can use these immediately to graft your scions onto, or if you wish to propagate even more rootstocks then wait until the following year when the rootstock has grown to around a metre tall.

In the UK grafting is normally done in early February, when the dormant season has ended and new growth is starting to appear. Don't leave it until the leaves have opened – that's too late unless you are pretty experienced. Here in southern France I have tried from early January to early March. The problem is that whilst nights can still drop to -5 °C, the days can be sunny with temperatures of up to 16 °C. This naturally affects the flow of hormones in the sap of the rootstock and the ability of the scion and the rootstock to bond together, and I have found that the optimum time here is the end of February with the potted rootstocks being protected in a shady, frost-free place with no wind access.

To whip graft select your scion stock from a named variety. You are looking for straight, pencil thick cuttings taken from the new growth of the donor tree. It needs to be showing three buds on the length if possible. The rootstock is prepared by cutting (secateurs are best) the rootstock stem off around 12 cm from the top of the root-ball. Save this cutting in a cool place to re-use as your new rootstock material. Then with a clean sharp knife; I use a grafting knife but a sharp Stanley knife or craft knife is also suitable, cut a 2 cm long vertical flap in the bark downwards, leaving the bottom of the flap attached.

Prepare the scion by cutting the bottom at a slant; just under the bottom bud if possible, aiming to make the cut and exposed cambium layer the same size as the flap on the rootstock. Without touching the cut surface with your fingers, try the scion to the inside of the flap. It needs to fit together without any gaps. You may need to trim away the top of the rootstock flap to allow access. If it fits snugly; and this takes practice, you are ready to bind the scion tightly, forcing the cut surface of the scion to the exposed inner surface of the flap. Use a grafting tape – they are available on the internet and are pretty cheap. Tape the scion horizontally, winding the tape so that it covers from the bottom of the flap upwards, but positioning it so that any buds on the scion aren't covered. Then gently tie it off. The binding needs to be tight, and this can take a bit of practice. A good long slanting cut to both the rootstock and the scion are easier to tape together. You can try a V shaped graft cut, where the scion is cut like an inverted V exposing cut surfaces on two sides of the rootstock, and the scion is cut to fit onto this. Again the shapes need to match so you get a snug fit and again a deeply slanting V makes it easier to tightly secure the tape. If it's not tight it will not hold the two living edges together to allow them to graft together.

Label with the variety and if the rootstock is in a pot, place in a cool, shady area. You will know if your graft has been successful in around six weeks, when new leaves start to appear from the scion. If no leaves have emerged wait a little longer. If the scion has gone brown and shrivelled it may have died. If you want to check, simply take a sharp knife and rub the edge of the scion bark. If it's green, then it might just be slow, but is still alive. If brown then it's dead. Failure is NOT a disaster. The rootstock should sprout new leaves from below the cut and grafted area and you can try again in a month or wait till the following spring. In the temperate UK I got 80-90 % success; here in France with extremes of temperatures in the spring, I only get about 60 % successes. Don't put the potted grafted rootstocks into a polytunnel, but place in a semi-shady place.

Now, the tops of the original rootstock that you cut off to do the grafting can be prepared to grow on as replacement rootstocks. Cut the bottom 2 cm off, and then cut the length into 20 cm pieces, insert the bottom into a seedbed up to 10 cm deep. Water it and these should start to form their own roots over the coming months and create a new rootstock by the following February. Remember to keep them moist.

CHERRIES, BULLACES, PLUMS ETC

The plum family has many members including delicious cherries and downright challenging sloes. But as a family, they are extremely useful to make a range of alcoholic drinks. They all flower in early spring, all have a stone, and have cultivated as well as wild cousins. The UK is an excellent source of all these fruits as the climatic conditions are ideal for them.

Cherries

There are two main types of cherry – sweet and acid. Sweet cherries produce delicious fruits for eating fresh, and are usually grown as small trees or trained as fans against a sunny wall. Acid cherries are excellent for cooking and grow well in partial shade. Both Varieties can be used to make alcoholic drinks. Cherry trees are ornamental as well as productive, with pretty spring blossom and colourful autumn foliage. There are choices to suit all sizes of garden, including compact options for small plots and even containers.

Cherries are normally grafted, with the most common rootstock being the semi-vigorous *'Colt'* rootstock. This rootstock is suitable for larger trees or growing as a fan against a large wall. This rootstock restricts the size to between 6-8 m tall and wide. For smaller gardens and containers, both *'Gisela 5'* and *'Tabel'* restrict growth to 3-4 m, which is much more suitable.

The variety is grafted onto the upper part of the tree and again, you need to make choices. Sweet or sour, early or late, self-fertile or needing a second tree as a pollinator. All sour cherries are self-fertile, but not all sweet cherries are. Some nice early varieties (good if you are in the southern part of the UK or can avoid late frosts) include *'Merchant'*, *'Kordia'* and *'Summer Sun'*. None of these are self-fertile, so will require a second tree of a different variety to pollinate. Late summer varieties include *'Lapins'* and *'Sweetheart'* (both self fertile), and *'Stella'* (self fertile, but splits in wet conditions).

Trees are sold bare-rooted or in containers. Bare-rooted are supplied during the dormant season and must be planted without much delay. Trees in containers can be planted all year round, but establish better if planted in autumn or spring. Choose a warm, sheltered site that avoids frost if possible and deep, fertile soil that is slightly acidic (pH 6.5-6.7 is ideal). If you are planting in containers, then choose robust containers that are at least 45-50 cm wide and deep. Use a soil based or multi-purpose compost mixed with some grit for drainage. If you have chosen early flowering varieties, you can cover the trees at night to protect from frost, removing this in the morning to allow sun and pollinating insects in.

Fleece or netting are also useful to reduce (you'll never stop it completely) birds stealing or damaging the ripe fruit. Oh, and a word about the cherry fly. This delightful creature will fly amongst the fruit, depositing eggs inside the ripening cherries with her ovipositor, which hatch out inside into small white maggots. From the outside you may not even notice the minute hole that the fly has deposited her egg in. As I'm not a fan of consuming additional protein without being aware of it, I now pick my cherries just before they are fully ripe, and when I take them into the kitchen I submerge the cherries in a large bowl of salted water with a plate to weigh the fruit down for ten minutes or so, and this soon encourages any little 'visitors' to exit the cherry. Then I rinse the cherries to flush them and the salt away.

Pruning is undertaken in July or August to avoid silver leaf disease and bacterial canker. Personally I only remove dead or damaged wood, and very long shoots, keeping pruning light. Try and shape into an open goblet shape, or in a fan shape if you are training against a wall. An annual pruning makes the fruit easier to harvest, to protect from birds, and allows replacement young shoots to take over from older fruiting wood. A little every year is sufficient.

Bullaces

Bullaces, damsons, sloes and cherry plums are basically different varieties of the same family and demand the same cultivation. In the UK bullace and their relatives can be found in gardens and growing in the wild. These varieties of plum prefer temperate weather and have very hardy natures. They are also noted for their sharp flavour, and instead of being used as a dessert fruit are normally used in pies or jam or for creating wines or liqueurs. Care is minimal once you get your trees or bushes established. Plant them about a metre apart and when established they will start to sucker; so it's a great tree or bush for a secure, flowering and fruiting hedge, alongside elder and hazel.

They prefer sun or partial shade and an ordinary soil. The tree develops long, downward growing branches when young which twist upward as they mature. They are self-fertile, with white flowers early in spring, producing a slightly oval blue-black fruit with a powdery bloom like sloes. The flesh inside is greenish. The plants should crop well in around four to five years, cropping in autumn; usually late September to early October. Any pruning should be done in the dormant season.

Plums

A plum is a good small tree and suitable for a garden or can be grown in a container. The plum family is huge including sweet and cooking varieties, bullaces, sloes, mirabelles and gages. They all like the same soil and growing conditions, and very little aftercare. Choose your rootstock as carefully as you would the variety. Plums are grafted onto two rootstocks - Semi-dwarfing St Julian A (grows to 4.5 m), and Pixy, a dwarf rootstock which grows to 3 m (perfect for a large pot or very small garden). If you have the space, most experts suggest the St Julian A rootstock.

If any pruning is required, do this in the dormant season, unless you are in a location susceptible to silverleaf, where the advice is to prune in late summer. Here in France commercial plum growing for drying into prunes is big business, and they prune in winter; so I follow suit and have not had any issues or disease. Planting your trees where they get plenty of light and air movement and this will reduce the likelihood of silverleaf disease, which is an airborne fungus. If you are troubled with this, rake up and burn all leaves, windfalls and prunings.

Some varieties of sweet or cooking plums and mirabelles may

have poor harvests every second year; this is perfectly normal, but choosing your variety carefully can avoid this. Overloading of fruit on the tree can result in branches splitting or breaking off, so reduce the load early in summer to prevent this by thinning the fruit or removing some altogether. Keeping branches short gives them more strength to carry their fruit.

Some popular eating varieties include '*Victoria*' and '*Marjorie's Seedling*'; both are self-fertile. The Golden Mirabelle '*Golden Sphere*', produces golden coloured little sweet globes in July which are great for cooking and eating. It's partly self-fertile. All are excellent for making wine or adding to liqueurs.

ELDER

Elder is one of my favourite plants and here in France, where it's dismissed as a weed, I have had to propagate my own in my roadside hedge next to my orchard. In late spring umbels of scented white flowers fill the air with a sweet smell, provide loads of pollen for the bees and then in summer the luscious dark blue-black berries appear, both to feed the birds and to provide me with a crop for making cordials and wine. The berries must be cooked or fermented, do not eat them raw.

Common elder (*Sambucus nigra*) is indecently easy to grow and requires little care. It can be found in waste ground and car parks, as well as countryside, parks and gardens. You can also get dark leaved varieties with pink flowers if you wish to make a pretty impact in your garden and they are still perfectly safe to use. Some great varieties include '*Sutherland Gold*' (yellowish foliage; if you use the berries, strain out the seeds and discard); and '*Black Lace*' (pink flowers and almost black leaves that turn red in autumn). The pink flowered varieties add a pink tinge to elderflower fizz and elderflower wine.

If you fancy growing elder, remember they prefer moist soil, so are ideal for that low-lying damp corner or near to streams, but they will grow perfectly well in hedges too. The tree can grow a little untidily, so you can cut it back hard in spring to maintain its shape and height. If you aim to collect the flowers or fruit a smaller tree is easier to harvest from.

You can propagate your own elder from hardwood cuttings in late autumn. Select your donor tree and cut pencil thick (or thicker) straight cuttings from mature wood (bark should be brown, not green) about 20-30 cm long. Cut the top of the cutting straight across and the bottom slanted (to remind you which way is up when planting, and the slanted bottom cut exposes more of the cambium layer for rooting). You can dip the bottom into rooting powder, but really, they are fairly easy to propagate, so it's not essential. Make a 10-15 cm deep hole with a metal rod or stick; or a slit trench with a spade, drop the cutting in and firm in and water. I tend to take more cuttings than I need to make up for any failures, but you should see new growth in spring. If planting in a hedge you may wish to protect with a rabbit guard until established and keep the grass short around the area. Water well in dry periods until growth is established.

You can also root cuttings in water in a jar on a windowsill, just remember to remove all the lower leaves, but leave at least two top leaves. Change the water every couple of days. Roots should form within a month, but allow a good network of roots before you pot up.

FIGS

In the UK figs may take a bit of time to ripen, but you can encourage this by growing against a sunny south-facing wall. I've seen fig trees laden with fruit as far North as Aberdeen, although they did have a bit of protection in the spring. The best varieties for growing in the UK are *'Green Turkey'* and *'Brunswick'*. Don't plant too close to a wall as the roots start to wander and may undermine the foundations, and if you plant and place a few old concrete paving slabs inside the planting hole you will restrict the growth somewhat, and channel the energy of the plant into fruit production. The fruit is ripe when it darkens and feels slightly squigy to touch. Beware the white sticky sap when harvesting – it can cause a skin reaction in some people.

The tree is deciduous and will drop it leaves in autumn, perhaps revealing some unripe fruit, which you can simply remove. Prune back in winter if the plant starts to outgrow its space, but expect an established tree to reach 3 m in spread. Against a wall, remove any growth that exceeds 500 cm outwards, and remember that you'll have to reach the fruit in order to pick them. Any leaves affected by mildew or insect damage can simply be removed, and as figs will happily root themselves from any branches touching the ground, remove all these and prevent the tree becoming a pest. Gather up any ripe fruit that had dropped from the tree to avoid wasps and to stop the tree seeding itself where you don't want it.

GRAPES

Grapes are perhaps the most obvious fruit to use for wine making, but unless you live in the wine-growing regions of France, Italy, California or Australasia, you may be daunted by the prospect. Don't be – even in the UK grapes can and are grown to make delicious wines and the process isn't hard. They are a versatile plant adding beauty and colour to a garden and add height, so can be incorporated along a fence or trellis or even grown in pots if you are short of space. In autumn the leaves often turn a pretty autumnal colour and the young leaves can even be harvested for use in Mediterranean and Middle Eastern dishes such as *dolmades*. You can even make a wine from the young leaves (just substitute them in the oak leaf wine recipe in this book). They are vigorous growers and once established will produce fruit for 30 years or more. Black grapes are normally used to make red or rosé wines, but you can make white wine from them – think about champagne; in additional to '*Chardonnay*' grapes it also contains '*Pinot Noir*' grapes.

For the average garden there are three main types of grapes to consider:

European (*Vitus vinifera*), American (*Vitus labrusca*), and these can be further split into suitable varieties for wine making. European grapes are better for wine production and are suited to warm, dry zones, and generally have a larger range of flavours than American varieties. In the UK grapes grown for wine can be grown outdoors, on a south facing wall or fence in the milder warmer south of the country, but will crop better in a greenhouse or polytunnel. When choosing your site avoid frost pockets, which can damage the new shoots, try and face south or south west and a minimum spring temperature of 16 °C. They come in white, red and black varieties, seeded or seedless, with different levels of hardiness and disease resistance. So, here are a few that do particularly well in the UK: '*Chardonnay*'; white grape, self fertile and hardy and can be used to make a still or sparkling wine (champagne is made from Chardonnay grapes), '*Dornfelder*'; black grape, self fertile, makes excellent fruity red wine with good disease resistance. '*Madeleine Angevine*'; self fertile, very hardy and makes an attractive fruity wine with a flowery nose, similar to an Alsatian Pinot blanc. And finally, '*Solaris*'; self fertile, hardy, good disease resistance and makes a nice fruity white wine as far north as Sweden! Some varieties, such as '*Muscat of Alexandria*', do better with additional heat, even in a greenhouse, in spring (to aid growth) and in autumn (to aid fruit ripening). Check if your chosen variety is self fertile or will need a pollinator – if you have only room for one plant this is crucial!

American grapes are the most cold-hardy, and a couple of varieties to look out for include '*Edelweiss*'; an early white grape for wine and eating which is hardy in zones 4-7 (up to -28 °C), and '*Seibel*', a group of hybrid grape varieties suitable for wine making which are also cold hardy and also grown in Europe, Australasia and the UK.

Choose vigorous one year old virus-free plants if possible, or if you have a friend with a healthy named variety you can save money and try some hardwood cuttings in late autumn or semi-ripe cuttings in early summer. Named varieties will not come true from seed, so this isn't really an option for the dedicated grape wine maker. If buying from a supplier or garden centre you may have the option of potted vines or dormant bare rooted vines. When choosing your variety, do consider your geographical location. Even in Northern Britain, sheltered Western areas, especially those benefiting from the Gulf Stream or a sheltered south facing wall can grow decent grapes – and remember global warming will only expand the areas in the UK suitable for cultivation.

Free draining soil on a slight slope facing south is the ideal location, but of course, most of us have to make do with what we have. If you have heavy soil, you can dig in lots of gravel to improve drainage. If you do not have the luxury of a garden, large pots filled with soil-based compost (John Innes No 3 is ideal) can be used for most varieties. Leaving the pots outside in winter will give them the dormancy they need. In polytunnels and greenhouses, ensure you have free draining soil and some humus mixed in to aid water retention in dry summer spells.

They like slightly alkaline or chalky soil, so if you have an acidic soil, dig in some lime or chalk to raise the pH.

So, you've chosen your variety and chosen your site. Grapevines are vigorous, and need trained and pruned. As vines can grow large and are not self-supporting you'll need to decide what support system you are going to use. There are three main ones:

Rod and spur or Cordon system

This system is usually reserved for indoor grapes or for growing against supporting walls. Fruiting side shoots are encouraged from a main vertical stem, a bit like an espalier. Use galvanised wire (2-3 mm diameter) stretched horizontally, spaced about 30 cm apart, using vine eyes to screw into wood or brick or to clip into the glazing bars of a greenhouse.

The Guyot System

This form of training and support has a single main stem which branches into one or two arms with fruiting shoots. It is mainly used for outdoor wine grape production. Two metre tall posts, 4 metres apart are driven about 45 cm into the ground and galvanised wires fixed with vine eyes screwed onto the posts at heights of 40 cm, 55 cm and then if required at 30 cm intervals.

Lollipop or standard system

Vines grown in pots or where space is limited are pruned and trained as standards with one main stem and a rounded head. Train the main stem up a stout bamboo cane and remove any secondary stems appearing from the base. Allow side branches to develop for the first two years; then in the third winter, remove all the lower side branches leaving only the top five or six branches. Prune these remaining side branches back to five leaves (winter) and pinch out any off-shoots on the side branches to one leaf (summer). Only allow one bunch of grapes to develop for this first cropping year. In the following years allow one bunch of grapes to each side branch. In early winter prune the side branches back to two buds when dormant. Pinching out is done in summer.

Planting should be done in the dormant season when the ground is not frosty or waterlogged, and the vines planted to the same depth as they were in the pots or if you've bought bare-rooted plants to the old soil mark on the stem. Soak bare-rooted vines for at least an hour prior to planting. In very cold areas, wait till March. Gently tease out the roots so they are well spread in the well prepared planting hole. If you have poor soil you can add a little general purpose fertilizer or compost. If planting outdoors against a wall, position the plants at least 12 cm from the wall, and at least 1.2 m apart. In open ground space the vines 1.2 m apart.

If you are growing 'inside' a greenhouse or polytunnel, you've got two choices. Greenhouse vines with their roots *inside* the greenhouse need more frequent watering than vines with their roots outside. Greenhouse vines perform best when the roots are planted outside and the vine is trained inside (through a gap at ground level). This isn't always possible, although could be easier with a polytunnel; but if you cant do this, then planting directly in the greenhouse border is the answer, but will require substantially more watering. One plant per average greenhouse is plenty, but in a large polytunnel you could space them at least 1.2 m apart. The best place inside the greenhouse or polytunnel is at the end opposite the door; then you can train the vine horizontally along the sides.

Water the newly planted vines when first planted and keep them watered regularly in the first year. If you see signs of powdery mildew on the leaves start to water well every ten days during the growing season. With outdoor vines you can mulch around the stem with gravel or stones. This will conserve moisture, reflect light back onto the lower leaves of the plant and deter weeds. In a greenhouse or polytunnel a mulch of manure just before the growth starts in spring, and a mulch of dry straw in summer will help to keep the atmosphere dry. Don't fertilize the soil in the first year unless your soil is really poor. A light fertilizing in the early spring of your second year with a high potassium fertilizer such as Vitax Q4 or tomato feed is beneficial. In indoor grown vines you should aim to feed every two weeks in growing season, stopping when the grapes start to colour up. Remove the tendrils as they appear. These are the natural appendages the vine would use in the wild to climb, but you are going to train your vines and the tendrils get in the way and use up the plants vigour.

All vines like good ventilation. It prevents diseases and helps the fruit ripen. In summer keep the vents of greenhouse grown vines open to prevent a humid atmosphere. You can close these as temperatures drop in autumn, but remember to open them again in winter to allow the plants to become dormant. It's also something to consider when it comes to pollination. No pollination means no fruit, so if your vines are greenhouse or polytunnel grown, open the vents fully at noon on dry days and shakes the stems vigorously to move the pollen amongst the flowers or use a feather or woolly dusting wand to transfer pollen from one bunch of flowers to another. This is especially good for Muscat type varieties. So summer is here, the grapes are swelling and colouring up and you are getting excited. Obviously geographical locations and the weather play an important part in ripening, but by late summer you should be seeing your grapes starting to look as if they are ready to harvest. If some of the grapes aren't getting enough light then remove a few leaves from around the trusses to allow the sun to get in. They are ready for picking when they are richly coloured, plump and taste sweet. Cut the whole stalk to remove and remember once removed they will not ripen further. You can store them on dry straw in boxes for a few days until you have enough to process. Protect your crop. If the birds have been stealing them you can net them, or if after all that not-so-hard work growing your delicious grapes you are visited by wasps or hornets, you can mesh or fleece them.

PEARS

Biting into a succulent, perfectly-ripe pear is one of the joys of autumn. You may be lucky enough to have one in your garden already that someone else has planted, but if not, they are easy to establish - you can even grow them in containers. Of course these are not the pears to grow if you want to make perry from them. That's a completely different kind of pear, but you can make a little juice from them to ferment as pear wine or to add to other fruit or floral wines.

A perry pear is normally hard and woody, growing on a large imposing tree that can take years to establish and fruit. There was an old saying regarding perry pears which was 'plant pears for your heirs'; but modern rootstocks mean that many of the old perry varieties have been successfully grafted onto modern, less vigorous and controllable rootstocks, which then fruit much earlier than their old predecessors. So, firstly choose the right rootstock, and then the right variety. The two main rootstocks are Quince A, common in garden centres and specialist suppliers and can be used for standards, bushes, wall grown and espaliers. Quince C is slightly less vigorous and suitable for containers, cordons, fan trained and espalier forms.

There are many old varieties out there – some grow tall, some medium height, and like cider apples they come in a variety of tastes. For example, *'Blakeney Red'* is a great variety for making medium-sharp perry; *'Barnet'* makes nice low-alcohol perry with a medium-sweet taste. Most varieties are not self-fertile and require a second pollinator.

Plant them the same as you would an apple, giving them a bit of space if they are grown on Quince A, and in the soil. Good soil, free draining and yet moisture retentive in a sunny spot with some shelter from wind. This is important. Bees are the main pollinator, and honeybees will not fly in strong winds. No bees, no pollination, no pears. Once they are established they need very little care. Pruning annually to remove dead, diseased and rubbing wood is key; remembering that the perry pear is a naturally upright growing tree. For those perry pears grown in containers, you will need to water well, especially when the fruits are starting to swell, and a high potassium feed is needed annually in spring.

SOFT FRUITS

This section contains fruits such as raspberries, strawberries and currants. These are easy to grow in all parts of the UK and most temperate countries. They are ideal to collect in small quantities and freeze to be used to make wines or liqueurs when you have collected sufficient amounts, and the left-over boozy fruit are great for crumbles and as a dessert with ice cream or crème fraîche.

Raspberries, loganberries and tayberries are easy to grow and delicious. Raspberries are either summer fruiting or autumn fruiting. This always confused me when it came to pruning them. Summer fruiting varieties need pruning immediately after harvesting - cut the canes right down to the ground. Leave the strong new young canes, but take out any spindly ones. Autumn fruiting varieties need the old canes removed in February. You can buy them as bare rooted plants or potted, and they can be grown in rows in the garden or in containers.

They like well-drained, slightly acidic soil. It's rather warm here in South West France for mine and they are resisting becoming established, but I'm persevering. Some good varieties include *'Autumn Bliss'* (autumn fruiting, good disease resistance) and *'Glen Ample'* (summer fruiting, good disease resistance and no spines).

A tayberry is a cross between a raspberry and a blackberry, and will give a good crop by the end of August in the second year. They too come in summer and autumn fruiting varieties. They freeze well and having a more tart taste than a raspberry are excellent for making whisky liqueurs. A loganberry is also a cross between a raspberry and a blackberry, but the plants are prone to spreading and the fruits have sharper taste than tayberries.

Strawberries are a wonderful early summer fruit, usually reserved for eating raw, but the sharper varieties and alpine strawberries are good for making Schnapps type liqueurs. Whilst large strawberries like fertile soil and full sun, alpine strawberries enjoy full or partial shade and, it is argued have a superior flavour. Whilst the alpine strawberries can be seed sown or transplanted into shady banks or slopes, traditional strawberries are planted as new young plants in September or April in beds or containers, mulched with straw and netted against birds, and you can harvest your fruit in June or July. Pinch out runners unless you wish to propagate new plants, and replace the old plants after three years. The varieties *'Gariguette'* (early); *'Hapil'* (mid); *and 'Mara de Bois'* (late; large fruit but taste of alpine strawberry) are all excellent.

Currants come in black, red and white varieties and their little strings of pearl-like fruits are easy to harvest from upright bushes. Blackcurrants are rich in vitamin C and have a tart flavour, excellent for liqueurs, wine and even *cassis*. Good varieties are *'Ben Conan'* and *'Big Ben'*, both can be grown in containers or in beds. Red currant varieties include *'Stanza'* and *'Red Lake'*. White currants have a more tart flavour than red currants. All currants can be propagated by hardwood cuttings from mid-autumn to winter. Prune out old wood in the winter to encourage new fruiting stems and harvest the trusses of fruit in early and mid-summer. They make an excellent wine for drinking on their own or when added to floral wines to add some body.

VEGETABLES

There are some vegetables which readily lend themselves to producing wines, allowing the self-sufficient gardener or canny shopper to take advantage of gluts and special offers. The more common vegetable wines include potato, rhubarb, parsnip, beetroot, carrot and peapod. They all have flavour and aroma, but need more sugar and additives to make a balanced wine. You can also mix and match, and even try something a little different by perhaps adding chilli or coffee beans in a recipe.

You can grow them in open soil or in containers. Potatoes can even be grown in old plastic compost bags, and simply upturned and emptied when you wish to harvest them. Most vegetables prefer a slightly alkaline soil, and some need protection from pests. I grow my carrots and parsnips in raised beds and protected with fleece over hoops against carrot fly. I reuse the homemade hoops and fleece every year and put them in place as soon as I've sown the seed. Once you get into the habit, it's not a chore. If you want to learn more about gardening for vegetables, I can recommend Carol Klein's book, *Grow your own Veg*, and all of Charles Dowding's books, Youtube videos and his website.

Freezing gluts from the garden in carefully labelled bags is now the norm in this house. There is nothing like taking what you *think* is a bag of chopped apples out to make an apple crumble or apple wine only to discover when it's defrosted that it is in fact parsnip chunks. Then I just sigh; put memories of the crumble to the back of my mind and start a demijohn of parsnip wine.

I'm a keen no-dig enthusiast with two large raised beds for my vegetables and fruit. I still weed by hand, and occasionally fall behind and let things slide in very hot days here in France, but as soon as the temperatures cool, I'm back out there, clearing a foot at a time. I only grow things that are expensive to buy in the shops or that I use a lot of. Mixed salad leaves take up a lot of space, as do winter squashes as I use a lot of these. I don't grow potatoes and have cut back on onions as they are plentiful and cheap here; but I do grow small pickling onions as I just can't find them and it is an essential treat in life for me to have a jar of pickled onions in my store cupboard.

No Dig Cultivation

If you'd like to try a no-dig bed, it's pretty easy and cheap; but you do need time. Mow very close to the ground your chosen site. Starting at a dormant period such as mid December is ideal. Mark out the area with a spade or trickle of sand or pegs and string and then cover the area with a thick layer of dense mulch. I've used a double layer of cardboard packaging, but you can use old woollen blankets or mulch mats.

Cover this layer with a deep layer of compost. Now, in raised beds this is easier as the walls will hold the compost in place, but in the open ground you might want to use railway sleepers or bricks to hold in the compost. Cover at least 15 cm deep and then place some more layers of cardboard on the top to exclude light and deter any weed seeds from developing.

Excluding the light with the cardboard and compost stops the weeds from coming through, although you will always get the odd perennial weed like dandelions trying their luck. If you see any green growth simply pull them up and remove, and then top up the layer of compost. Eventually the weeds and grass weaken and give up. Here in France I've had a problem with bindweed in the raised beds I inherited from the last property owners. They were not gardeners and had left the beds uncovered and the bindweed was rampant. It took me a full nine months of almost daily weeding and digging out to finally rid myself of bindweed without the use of glyphosate weedkiller.

The cardboard will eventually rot down and like the compost will be absorbed into the underlying soil, raising fertility as it does so. Keep topping up the compost layer, keep removing any unwanted green growth and then when clear you are ready to plant up.

If you wish to just have a traditional, diggable plot or are sited at an allotment, then remove all perennial weeds, incorporate as much compost or manure as you can access and then cover until you are ready to plant out. Think about a simple crop rotation plan and also about producing your own compost. I try and rotate the beds every year, simply moving the groups of vegetable up like a conveyor belt, knocking the top one off to begin again at the bottom of the bed. The bottom half of the bed is manured every winter, whilst the rest gets a light dressing with home-made compost from the composter.

CROP ROTATION

CROP ROTATION PLAN

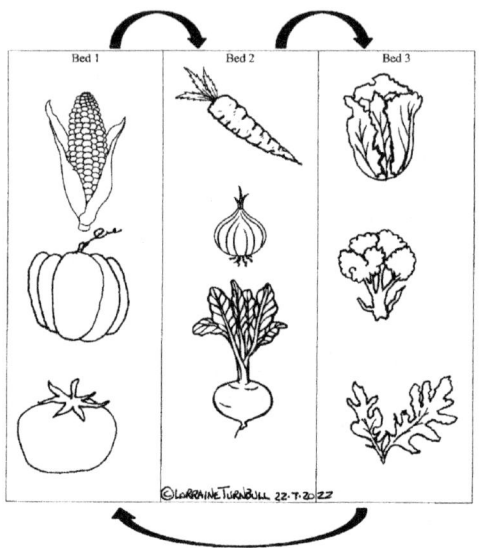

Crop rotation is an old gardening practice where certain groups of vegetables are moved as they use the soil. A simple crop rotation plan is over three year, where if plants are grown in a bed, they are simply rotated in order to avoid them growing on the same spot every year. This reduces the chance of diseases in the soil, of a build up of pests, and allows for greedy plants to be moved to fresh soil.

Crop rotation is mainly used for vegetable growing, as most fruits are permanent plants that cannot be moved, so I keep my rhubarb in the same place in the plot permanently, only lifting and dividing every five years and discarding the old parts and replanting the new stool at the opposite end of the bed in richly prepared soil. My currants and strawberries are permanent members of my dedicated fruit bed – they just get a good topping up of compost in winter.

You can see the crop rotation plan opposite and may be wondering where you add the new compost or manure; well, in winter you add this to the place you are always going to grow the greedy feeders. As you rotate the plant groups backward, you will always have the greedy feeders making the most of the new manure or compost, and the medium feeders following them. So In year 1 the greedy feeders are in the first bed, then in year 2 you move along a space and plant the greedy feeders in bed 2, and then the next year the greedy feeders are in bed 3.

Greedy feeders include tomatoes, sweetcorn, squash, gourds and peppers.

Medium feeders include roots & legumes – onions, carrots, parsnips, beans, peas and beetroot.

Light feeders include salad leaves, broccoli, cabbages and other brassicas.

Plant propagation

There are three main methods of producing plants - from seeds, from division and/or layering and from cuttings or grafting.

Commercial seeds can be bought almost everywhere and the seeds inside the packet will normally grow to resemble the seeds on the picture of the pack. Self-harvested seeds may differ slightly from the parent plant because plants will hybridise in discriminatively and may not grow true to type.

In reality this can be an issue if you save your own seeds of plants from the gourd family, including cucumbers, courgettes, melons, marrows, squash and courgettes. Commercial seed are specially bred for low levels of the toxin *cucurbitacin*, but if you grow fruits from your self-harvested seeds the fruit may be toxic. The symptoms are stomach ache, nausea, diarrhoea.

One annoying thing about seed packets is that they don't always tell you how deep to plant the seeds (or in the flower garden bulbs). A good guide is twice the diameter of the seed (or bulb). So if you are planting peas then plant them two times as deep as they are in diameter. The bigger the seed the deeper you plant.

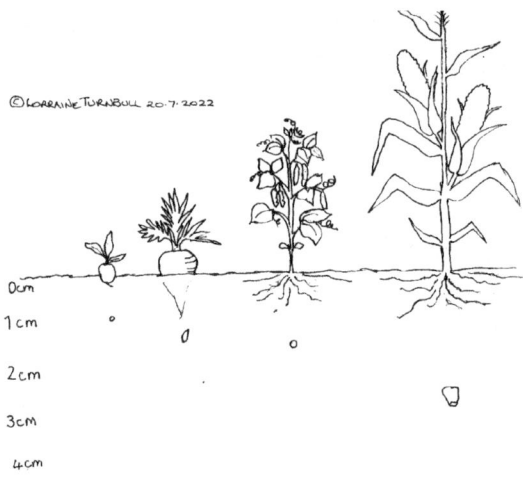

Division is usually done to increase herbaceous flowering stock, not vegetables. So, for example, to increase your stock of daylilies, you would fork out a clump (either early spring or autumn are the right times) and divide with a sharp spade into good size clumps and then replant. You can use this method when lifting an old rhubarb plant, splitting it to ensure that each new piece has an 'eye' and then replant in well fertilised soil. Some plants can be divided by hand by teasing out the individual plants in a clump.

Layering is mainly done where plants have long, lax growths that when they naturally touch the ground they produce roots and thus propagate the parent plant. You can mimic this behaviour by bending a suitable stem to the ground and securing it with a tent peg or large heavy stone, ensuring the soil covers the parent stem. This should develop roots in a few months, and you can then gently cut the new rooted baby from the parent plant, leaving it for a further month before lifting it to move elsewhere. Some species take longer than others, but good species to try this with include blackberries, brambles, strawberries and roses.

Cuttings can be propagated in spring, summer or winter. I personally prefer to take hardwood cuttings of most species in late autumn, after leaf fall, in a seed bed and checking if they have been successful the following early summer. I always take more than I need and at least some, if not all are successful. Roses, flowering and fruiting shrubs such as currants, and even vines can be propagated this way.

Grafting and budding are ways to attach a named variety onto a rootstock using a small twig (scion) or bud inserted into the rootstock bark. I explain how to do this in the section on growing apples, but it is equally relevant to other tree fruit, but you must keep the families separate - that is you can only graft apples to apples, pears to pears, and for stone fruit, such as cherries, plums, peaches and nectarines to St Julian A or Pixy rootstock.

MISCELLANEOUS

Nettles

I don't think anyone actually needs to cultivate nettles as they are so common everywhere in the UK, but I do have a patch in my wildlife garden for the butterflies, and they are immensely useful. The young tops can be used to make nettle beer or wine, but I also half-fill a plastic dustbin with lid with nettle tops and top it right up with water. In a few weeks it has transformed into a strong fertiliser which can be diluted down one part liquid to nine parts water and used as a feed for most plants including fruit and vegetables. I do have to warn you the smell is strong and horrible; don't keep your container near the house. Cutting off the old nettle tops and using them like this will encourage lush new growth should you wish to use the young nettle tops after midsummer.

Hops

The hop plant, *Humulus lupulus* is a perennial herbaceous bine, and is famous for adding the bitterness to beer. It self-twines around a support, produces its flowers in summer, and, if you have successfully chosen a *female* plant produces thick, sweet-smelling papery cones. In winter, the plant dies back, to reappear the following spring.

Hops grow best in the far south of the UK, and can be found growing wild in many places in southern and south western Britain. They are also produced in Germany and areas of France. They need full sun to produce quality cones. Many factors including soil structure, acidity and mineral content, as well as climatic variations of sunlight and rainfall will affect the taste of your hops, meaning that the same hop variety grown in different geographical locations will impart different flavours when making beers.

The simplest was to cultivate is to buy a rhizome, either in a pot (rooted) or dry rhizome (unrooted), and they are widely available from specialist growers or home brew suppliers. Plant out in spring, after any risk of frost in a deep, moisture rich slightly acid soil where you can erect a trellis or other tall support for them to scramble up. Water frequently until the plant shows it has established and thrown up new shoots.

There are many varieties, including golden-leaved ones to add a little drama to your garden as well as providing you with a crop of hops for brewing. *Humulus lupulus 'Aureus'* (the Golden Hop) can grow to an enormous 6 m, but *Humulus lupulus 'Golden Tassels'* restricts its growth to around 3 m. A couple of other classic home-brew favourite hop varieties include '*Fuggle*' (minty & floral), '*Goldings*' (spicy & honey flavour), '*Target*' (citrus & spicy), and '*Wye Challenger*' (spicy & green tea). Remember only the female plant produces the cones.

Once chosen and planted you'll need to quickly erect a sturdy vertical trellis or a few tall poles with hanging vertical lines or strings, securely embedded in the ground. Hops can grow up to 8 m tall, and when ready to harvest can weigh 12 kg. They will self twine around any support, so you do not need to tie them in. Avoid synthetic strings or lines which have a slippy surface for the bines.

Keep the ground around the base of the plant weed free and give the occasional feed. Remove weak bines at ground level, and remove the bottom leaves to help keep the plant pest free and easy to weed. The cones will be ready to harvest by early September, when they feel dry and papery, have a strong '*hoppy*' smell and leave a yellow powder to the touch. The hop cones may mature at different rates, so check every few days, starting at the top and working down the plant. Dry in a cool airy place by suspending from their strings horizontally if possible – an open ended polytunnel or large shed is ideal. Once dry, seal them in a zip-lock or similar bag, excluding as much oxygen as possible and freeze.

After harvest, cut the remaining bines down to 60 cm from the ground. Once the frost has been, cover the location of the plant with a mulch and or fertilizer. If you have had a very good harvest, some of the dried cones can be placed in a fine mesh or linen bag and placed under the pillow to aid sleep, as hops are in fact part of the same family as the cannabis plant.

Flowers

There are many wild and garden flowers that can be used to make pleasant wines or as an infusion. I'm not going to detail the process of growing them here as that subject is vast. Instead, I will list some of the more common and reliable ones that are non-toxic. Honeysuckle, roses, dandelions, coltsfoot, burdock, heather are just some that make nice wines. Thyme, lavender and rosemary have quite strong tastes, but lemon balm is a good option; and experimentation is the name of the game, so if you fancy it; by all means give it a go by exchanging the main ingredient, just remember to reduce the quantity as these herbs have a concentrated taste. In France gentians are used to produce a spirit drunk mainly as an aperitif, but having tasted this I'm afraid I have to say it's an acquired taste and not really for me.

Honeysuckle will happily scramble through a hedge or along a semi-shaded fence. Some varieties are strongly scented, some have no scent at all, but hey make a nice light floral wine that can be given some more body by adding either grape juice concentrate when making or by blending in some pear wine or rhubarb wine when bottling. Roses can be ground cover, bush, ramblers or climbers, and some flower once in early summer, whilst some are perpetuals that will flower on and off all summer into late autumn. They come in a wide range of colours and scents, and most are easy to grow and propagate as hardwood cuttings. For flavouring wines or liqueurs try to collect petals from the stronger scented types, such as musk, province or rugosa roses.

Dandelions need no cultivation and will of course pop up everywhere from early spring. It's a really versatile plant and can be used for a variety of culinary purposes as well as wine and beer. Heather comes in two main families and can be summer or winter flowering. They like a sunny spot and, although some will tolerate a little lime in the soil, they prefer deep, acidic soil. If you want the flowers for ale or wine, choose one of the larger flowering varieties to grow. Bell heather (Erica cinerea) is the wild summer-flowering heather you can find growing wild on heaths and sea cliffs all over the UK, or can buy from most garden centres. It likes acidic dry and well drained soil. Heather tops are the flowering tips, and can be used in heather ale and metheglins.

Herbs can be grown in pots and window boxes as well as in sharply draining soil in a garden. They like full sun and hate damp spots. Even if you feel that the strong herby scents are to be avoided in wine making, they are pleasant in some liqueurs in small quantities.

Ginger

Ginger is a tropical plant and if you live in a cool climate such as the UK, will only really be happy in your home, where it could be grown on a south facing kitchen or bathroom windowsill in a pot. It is a root spice, and to grow it just buy a small piece from a supermarket. You're looking for a firm, plump root with one or two 'eyes' visible. These will become the new stems of your ginger plant. Fill a pot with a 70-30 mix of compost and gravel, and push the piece of ginger into the mix till it lies horizontally and partly covered, with only the part where the eyes are located showing above the compost. Water, keep in a warm spot until new leaves emerge and wait for shoots and new roots to develop, then move the pot to a warm bright spot. Water regularly and your ginger plant should be ready to harvest in autumn. You can then cut a starter root with eyes again and after allowing the cut to callous over for a couple of weeks replant the new 'mother' ginger plant.

PART 2

BEER, CIDER AND WINE

BEER

From personal experience, I would suggest NOT using the airing cupboard in the home. A well insulated shed or garage is ideal, with temperatures between 20 and 35 °C. If you are going to struggle to maintain a constant temperature you can place the fermentation vessel on a regulated heat mat or fit it with a heated belt. Finished beer or lager must be stored in a cool, dry place.

Cleanliness, as in all booze making, is crucial. You must sterilise everything. Do not use bleach – it taints everything, instead use sterilising tablets used for baby feeding equipment or Campden tablets.

Beer kits

This is the standard method if you buy a DIY beer kit. Most beer kit instructions are the same, so you need to firstly READ the instructions, but I've summarized the general instructions here:

Heat the unopened can of malt by immersing in hot water. This softens the malt and makes it easier to pour out. Empty the can into the brewing bucket, and rinse the empty can with hot water and add this to the brewing bucket.

The best kits will not require you to add any sugar – again, read the instructions. Top the volume in the brewing bucket with warm water until the final temperature reaches around 25 °C. Stir well.

Add the sachet of yeast, stir gently and replace lid, fitting an airlock. Leave undisturbed in a warm place (20-25 °C) for between 5 and 7 days. Fermentation is slower in cooler temperatures. You do not want a ferociously bubbling airlock – this indicates the temperature is too high.

Use a hydrometer to check the S.G. When the reading is around 1.010 (dependent on the type of beer or lager you are making), and the bubbling has stopped, fermentation is complete. Syphon the beer into bottles or your beer barrel. This is the time to prime your beer as per instructions. You can also add hop extract or dried hops if you wish.

Seal the barrel and transfer to a warm place. The yeast will react with the priming sugar and result in carbonation (gas production). After 2-3 days move the barrel/bottles to a cold place to allow the beer to clear. The sediment will fall to the bottom during this time, so choose a permanent storage area where you will not need to move your beer; and a little tip – with the tap of the barrel facing forward, slide a small wedge or wedges under the front and this will encourage sediment to settle away from the tap. Clearing can take anything from a week to a month, bottles obviously clear quicker as they contain less volume.

Your beer is now ready to enjoy. You may have to draw off the first quarter pint to discard if it contains sediment that has settled in the tap (if you didn't use the little wedge tip). Pour gently from the barrel. You can fit a CO_2 gas canister to top up the gas level in the barrel and prevent spoilage by oxygen. Unopened, your beer should be safe for up to six months, and once open use within three weeks.

There are many recipes on the internet, and in this book and if you fancy making your own, then the recipe below should be a good starter to make around 24 pints.

> 50 g dried hops
> 750 g sugar
> 1 kg malt extract
> Sachet ale yeast (to make 24 pints)
> Water

1. Boil 6 litres of water in a big pan; add 50 g of hops and leave to boil for a further 30 minutes.

2. Slowly add the malt extract, stir well, and add the sugar, stirring well, and allow to boil for a further 5 minutes.

3. Cover a sterilised large fermentation bucket with a clean muslin cloth (you'll need to fasten this securely), and strain the contents of the pan into this carefully. Now is the time to carefully lift the

bucket and move to the place where you'll leave the bucket to ferment without disturbance.

4. Add a further 6 litres of cold water and leave the mixture to cool for 30 minutes.

5. Add the sachet of ale yeast and then cover the bucket with the lid and leave to stand in a cool, dry place to ferment for 10 days. You should now observe a crusty cap on the liquid, and can now prime your bottles with a small amount of sugar ready for bottling. I use 2.5 ml (or less if you can taste sweetness) per 500 ml bottle. Syphon the beer into the bottles, avoiding disturbing the sediment and the cap; filling to the bottom of the neck and cap with crown caps. Store in a cool place for at least 14 days.

Now, if you feel you'd like to move on from beer kits, you can do the whole thing from scratch or cut a lot of time out and buy a ready to brew mash extract (this will save you 3 hours of work).

There are loads of different styles of beer. To name a few - lager, bitter, IPA, stout, dark ale, blonde ale and wheat beer.

Brew in a bag (BIAB) has become incredibly popular because it lets you miss out the whole sparging process. I use a reusable brew bag for this. To begin the mashing process, put your chosen grains in the bag and put it into the hot water according to the recipe you have. Remember and gently stir the grains to allow them to release the sugar.

Check you maintain the temperature for the time your recipe says, giving the occasional stir. When complete, remove the brew bag from the liquor, lifting it out slowly (like a giant teabag) and letting the liquid fall back into the pot. Drain; don't squeeze. If your arm is going to get tired then you can hang the bag from a hook over the pot to catch the dripping liquid until it stops. Your liquid is now called a 'wort', and is now ready to boil. If you wish to add hops you do it during the boiling process. If you add the hops at the start of the boil, you will increase the bitterness of the beer, if you add at the end you will get more aroma. So bring the wort up to a rapid boil, then if you want add your hops. Boil the wort for 50 minutes or the time suggested in your recipe. If at 45 minutes you wish, you can add a little more hops to increase your finished beer aroma. Turn off the heat at 60 minutes and allow the wort to cool.

Now this can take ages, so you can fill a sink with cold water and place the pot in this which will cool it faster. If the sink water heats up from the pot, simply drain and refill the sink with fresh cold water. So once the wort is at room temperature you can transfer the wort to a fermentation vessel (through a strainer if you used actual hops and not pelleted hops), add the yeast, fit an airlock and place the fermenter in a cool dark place and wait 2 weeks for the beer to ferment. If you want to be able to calculate the ABV then take your initial S.G. reading when you pour the wort into the fermenting vessel.

In the last week of fermentation if you wish to add any fruity or spicy flavourings this is the time. Ensure you sanitise anything you add now or you'll risk the beer being contaminated.

When you think the fermentation is coming to an end move the fermenting vessel (if you can) to a really cold place to force the yeast and any sediment to fall to the bottom. If you can't move it you can use a temperature controlled jacket to wrap around the container to chill it. Now is the time to take your final S.G. reading and do your calculation to work out the ABV (Check the section on equipment to see how to do this).

You can rack off now into a clean container ready for bottling or if your beer has cleared by chilling just go straight to bottling with careful use of the syphon tube well above the lees. Your beer won't have much fizz, so you can prime your glass bottles now, prior to filling. See your recipe for priming amounts, but a good practice (and you really want to avoid making a bomb) is to use a quarter teaspoon or 3 g sugar per 500 ml bottle. Crown cap bottles or flip top bottles are perfect. Run the beer with a syphon tube into the bottle, fill to within 5 cm of the top of the neck, cap immediately and store in a cool place for 3 weeks, then in a cold place for at least a week before you drink.

LAGER

Lager is just variation within the larger beer family, which includes ales, bitter, wheat beers and lager. This isn't really a recipe, but is an explanation of how to make lager from a buy-off-the-shelf lager kit; a good introduction to the method. The kit will normally include malt extract (liquid), lager yeast and hops. You will also need 1 kg sugar for making the lager and some sugar to prime the bottles for carbonation.

Prepare an area where the fermenting brew can be undisturbed and is between 20 and 25 °C for 10 days. Pour the sugar into your fermentation bucket. Add 3 litres of boiling water and stir to dissolve. Pour the liquid malt extract into the bucket. If this comes in a tin, you may have to heat the tin in a pan of hot water to soften it. Stir well.

Fill the fermentation bucket up to the 23 litre line with cold water (or as per instructions of your own particular kit). Check the temperature of the liquid. It needs to be below 26 °C before you proceed.

Sprinkle the yeast on the surface. Do not stir. Fit the lid, but do not click it tight. Leave in a constant temperature for 10 days. On day seven, add the hops sachet, do not stir, and then gently replace the lid.

After the 10 days taste the lager. If it is still sweet and bubbling then further fermentation is required. Taste again in 2 days time. If satisfied with the reduced level of sweetness, then you are ready to bottle it. Prime the bottles and fill as you would for beer.

CIDER

Put simply, Cider (or hard cider, as is known in the USA) is the fermented juice of apples. It can be made from almost any type of apple; however you will get a more complex range of flavours and mouth-feel if you use a mix of apple types.

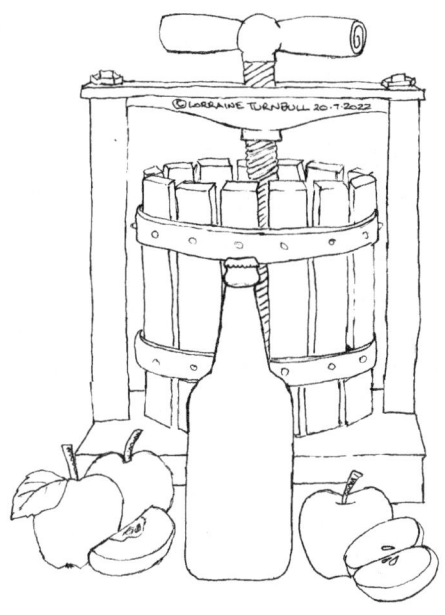

Apples are divided into four taste groups – Sweets, Sharps, Bittersweets and Bittersharps. Sweets are in the main dessert apples – the good old Cox and Discovery are sweets. Sharps include cooking apples such as 'Bramley' and crab apples. Bittersweets and Bittersharps are mainly found in cider apples, and they have a higher concentration of tannins.

So, a mix of mainly dessert fruit, a few cooking apples and if you are lucky enough to have them some Bittersharp and/ or Bittersweet cider apples will give you a really good mix for juicing and fermenting. This is fine if you have access to an orchard, but for the average gardener or someone without even a garden who is reliant on buying apples, you may have to search around. You can; and a lot of commercial cider makers do indeed make a single variety of cider, from 'Pink Lady' or 'Discovery'. You'll need to see what your particular taste buds enjoy and cut your cloth to suit.

For example in my house I like dry cider, my husband likes sweet and my son likes medium and sparkling. My best friend likes a traditional full bittersweet. One is not better than the other; it's just that we all have individual tastes and favourites.

When I used to run cider making courses, back in the day, people would often complain that they only had a 'Bramley' in their garden, and I'd advise making a mix by using 20 % 'Bramley' and buying in about 80 % dessert apples, or join a social media group and see if anyone locally had any cider trees or is willing to donate some dessert apples to you. There is a solution to every problem.

When the summer turns to autumn it's time to start preparing. Are you going to use a tabletop method and make a very small amount that you can fill a demijohn with? Or are you going to try hiring a scratter and press from a local apple group and spend a weekend picking, washing, scratting (pulping) and juicing? They are two different ways and scales of making cider and you need to decide a few months in advance. Many towns and villages in the UK have Apple Days where the equipment is already set up and you can take your apples to be pressed and collect the juice there and then. Check locally on the internet, but most are around the 25th October annually.

Naturally, it's not quite that easy, as apples ripen at different times according to their variety and the local climactic conditions. Desserts such as 'Discovery' and 'Gala' are usually the first, ripening at the end of August/start of September. Some old-fashioned dual purpose varieties and cider apples can ripen at the end of November. You can check if they are ripe, by taking one in your hand as it is growing on the tree. If it's ready to be picked it will 'give' a little when you lift and gently pull it. The skin (unless it's a russet with a rough skin) should feel slightly waxy, and when gently pressed with a thumb may 'give' a little. Finally if you cut the apple through the core a ripe apple will have brown pips, an apple that will be ripe in a week or so will have half brown and half white pips and an unripe apple will have white pips only.

WHY is this important? Well, apple juice contains a lot of sugar when the apples are ripe and it is this sugar that is turned into alcohol. Unripe apples are full of starch which doesn't ferment. You can store the apples for a week or so in a cool, dry and vermin proof shed to increase the sugar content. In a good summer sugar levels in apples might be as high as 17 %, but in a cool wet summer less than 10 %. Of course, commercial cider makers often add sugar to the juice to bring it up to the required level for fermentation – for them consistency is key (many also add a variety of colourings, flavourings and other sweeteners, but we wont go into that here).

You will also have heard various tales of old cider makers throwing in a rat, or adding sheep's blood or using fruit that is rotten and black. Let me dispel a few myths here. Perhaps, way back in medieval times some of this may have actually happened; there is a theory that cider fermentation can slow or stop unless it has the required protein, and perhaps this may have been the case back in uncivilised times; BUT nowadays you should be able to make cider cleanly, and if you have a problem with a slow fermentation (nothing actually wrong with that), then you can add nutrient to it in a measured dose (your home brew shop should have this in stock).

Please don't use mouldy fruit for cider, no matter how desperately short you may be for juice. The apples must be free from any cuts, damage or moulds. Basically, if you wouldn't take a bite out of it, get rid of it. They compost well.

Processing

So, in the first section of this book we've discussed apple varieties and types and we now move on to the processing. You will need to wash the apples prior to milling (scratting or pulping) them. I use a big polypropylene trough, but an old baby bath will do. You need to wash them because they may have mud, bird poo, slugs and insects on them. Change the water when it turns cloudy. Put the clean apples into drainage crates and stack next to the scratter for milling. Placing the wash bath on a stand will greatly avoid a sore back.

Milling

Whatever you call it – milling, scratting, pulping it's the same thing – breaking down whole apples into much smaller pieces of apples. If you want to make over 100 litres I'd suggest hiring a *Speidel* mill or *fruit shark*. Small quantities of apples can be processed with a food processor in the kitchen. Whatever you use, thorough cleaning afterwards is essential to prevent the acid from the apples eating into the metal blades. Try to mix sizes of apples, small crabs can clog a scratter, so mix them gradually with medium size apples, larger apples can be cut in two. The apples pass through the top of the scratter and fall into a bucket places below. Please use a plastic bucket, not metal. A sheet of plastic underneath the bucket will also keep the working area a little tidier. The pulp falls through the scratter pale green, or flecked with red if you are pulping red skinned apples; but it will quickly discolour to a dirty brown colour. This is caused by oxidation and is perfectly normal.

If you are using the equipment to make juice and NOT cider then you can add 5 g Ascorbic acid (Vitamin C) to the pulp (10 litres of pulp) as it falls through, mixing thoroughly and this will prevent the oxidisation to some extent, and give you a clearer apple juice. DO NOT add this to juice if you are going to make into cider.

Pressing

beSo you now have buckets full of pulpy apples. How you press is up to you depending on the scale of your operation and your finances. You can buy or hire (or make) a small table-top press and press enough juice to fill a demijohn, or you can buy or hire a large rack and cloth press or water bladder type press. The largest commercial operations tend to opt for a belt press or a double rack and cloth press. In olden days straw or Hessian sacks were used to separate layers of apple pulp and the wooden pressing boards were screwed down to force the juice out.

We use a rack and cloth press with a hydraulic jack, that John made basically copying a commercially manufactured modern small commercial *Voran* press. Our press cost us a fraction of the cost of buying one, but John is a welder fabricator and had the expertise, time and knowledge to be able to construct one. There are details in the appendix of how you can make a very basic rack and cloth press of your own using a normal car jack.

So, with a rack and cloth press, you place the former on the bottom juice collecting tray, and lay inside a press cloth, and in this cloth you fill the cloth with pulp up to the level of the former, then remove the former and fold the cloth over to make a closed package or 'cheese'. Then pop on a press board, and put the former back on top of this and begin again. A tower of five or six cheeses separated by boards is topped by one or two pressing boards and the pressure blocks and the hydraulics force the blocks slowly down, forcing the juice out of the cheeses, into the collecting tray and by hose to the collection bucket underneath. This is then either carried (or pumped) to the fermentation vessel.

The amount of juice you will produce is very dependent on variety, and climate conditions. For example, early dessert varieties always produce more juice than cider apples or russets, but a good mix of 25 kg of apples will produce around 16-18 litres of juice. If you don't have enough apples to produce enough juice to fill up to the shoulders of a demijohn, then go to your supermarket and buy a couple of cartons of pure apple juice and add this to top it up, just to the shoulders. Any fermentation vessel must have a bit of headspace to allow for the sometimes enthusiastic start of fermentation.

The next step is taking an initial Specific Gravity reading with your hydrometer, right after you've done your juicing. Now this sounds technical and complicated; but trust me it's quite easy, and again in the first section of the book, I explain how to take a reading with a hydrometer and how to do the calculation.

If you've got pretty ripe apples I'd imagine it would be anywhere around 1.070 (high sugar content) and if they are perhaps not all really ripe then maybe around 1.045. If the reading is showing less than 1.045 and you don't have any sweet juice to blend into the fermentation vessel then you can raise the level a little by adding sugar. This will prevent spoilage due to a very slow fermentation. To raise the level in 5 degree steps, check how many litres of juice you have in the fermenting vessel and dissolve 12 grams of sugar per litre. Pour the juice from the hydrometer back into the tank, stir to dissolve and mix the sugar and begin to take a new hydrometer reading.

So, you've tested again and your reading is S.G. 1.070 – this means that you can expect your final cider to come out at around 8.5 % ABV. If your reading is S.G. 1.045 then you have the potential to make cider of 6 %.

Remember that the aim is to make a nice palatable drink for quaffing. 8.5 % is pretty strong and quite hard to drink. Session ciders sold in pubs are usually between 3.8 and 4.5 % ABV. You can of course make stronger cider if you wish; or take a cider that's stronger than you like and dilute it down with apple juice and or a little water (if you pasteurise it).

So this initial S.G. reading is your starter measurement. Write down the date and the reading clearly in your little home-brew notebook. You'll forget otherwise.

Pre-Fermentation

I've added this in as it's optional, and I'll explain why. You've got your juice sitting there all nicely in its fermentation vessel, whether this is a demijohn, a 200 litre barrel or a 1000 litre food grade IBC container. You've taken the initial S.G. reading. Hurrah!

Easy peasy so far. Now, I've mentioned adding sugar. Let's talk about this a little more. As a small commercial cider maker, and now as an amateur maker and drinker, I don't need to bump up the sugar content in my juice. Many large commercial makers do this. They are creating and selling VOLUME. I always have and always will create a full juice cider made with at least 98 % full apple juice. There will always be a tiny amount of water, and a fraction of yeast and perhaps sodium metabisulphite, so it's never going to be 100 % juice, but it is almost.

If you want to make a really strong cider, by all means add extra sugar if you wish, but beware that you may cause the yeast to fail as the growing conditions are imbalanced.

Initially, I take my S.G. reading, mark it in my records and let my juice sit. I may take a pH reading to see how acidic my juice is, but this is not required unless you are really making a lot of cider and have a bit more experience.

Here's the science about acidity and pH -

The acidity is controlled more by the variety of fruit than the climate. As beginners we are just looking at pH, as it relates better to various aspects of fermentation biochemistry. Narrow range pH test strips (e.g. pH 2.8 to 4.2) are now available cheaply from most home brewing suppliers. A desirable juice pH range for cider-making is say 3.2 - 3.8. At higher pH the fermentation will be subject to microbial infection and at pH 4.0 or above this can lead to serious flavour problems. Many traditional bittersweet cider apples tend to be high in pH which is why they need blending with more acid fruit, preferably before fermentation. That is one reason why bittersharp apples, such as *Kingston Black*', have been regarded as near perfection in terms of their composition for single-variety cider making. If juice is too acidic you can raise this by adding a little calcium carbonate to neutralise it, in 1 gram per litre steps. Personally, I hardly ever test with pH papers and just ensure I have a good mix and rely on my taste buds to tell me if the juice is too sharp or not sharp enough.

You may also wish to add sodium metabisulphite or Campden

84

tablets to kill any natural yeasts in this juice. Again this is optional. Personally, I like to keep my cider making very clean. All my equipment is clean, apples are clean and my fermentation vessel clean before use. The apples and therefore the juice will naturally have some wild yeasts on them. This is fine and normal. If you wish to kill off the natural yeasts in the air, then this is the time to do it, when the juice is sitting in the fermentation tank, before you add any additional yeast to it. So, the day the juice is pressed is day 1. Sometime on this day you can add the Campden tablets or sodium metabisulphite. The formula to use is 10 g to 100 ml of water to make up a 5 % solution. Then add 1 ml of this per litre of juice. So if you have 100 litres of juice you need to add 100 ml of this solution. Mix well with a long stirrer and leave covered overnight. Note: Campden tablets are formulated with metabisulphite to give the equivalent of 50 ppm sulphur dioxide when each is dissolved in 1 gallon of liquid.

The Yeast

Stick to a good general purpose wine or cider yeast - not a brewer's or baker's yeast. Follow the yeast supplier's directions. If Campden tablets have been used, it's important to wait overnight before adding the yeast culture. Fermentation should commence within 48 hours if an active yeast culture is used. Wild yeasts (if you didn't use the Campden tablets) will take longer and will eventually compete with any additional yeast you add.

Fermentation

You may not see anything happening for a day or two and then you will see a seething and gentle bubbling or frothing. If your fermentation vessel has a wide opening, you can put in a loose wad of cotton wool into the neck, swapping this after a week or so with a proper airlock to allow the carbon dioxide gas to escape but not allowing any bacteria or oxygen to enter the fermentation vessel. Take this opportunity to top up the liquid level with more apple juice or a mix of apple juice and water, still leaving a little headspace. As your juice turns gradually into cider you must remember that **oxygen will ruin your cider and must be excluded**. This is why you need the airlock, to prevent any oxygen getting through to the cider, but allowing the carbon dioxide to gas off. As the fermentation slows down as your S.G. readings drop and begin to level off, you should consider racking off at an S.G. of 1.005.

You can perhaps skip the nutrients unless the fermentation begins to 'stick', (at above 1.005) or unless you know that your fruit comes from big old trees with very low nutrient levels and you are not prepared to wait a few months. Apple juices are generally low in yeast nutrients so your fermentation rate will probably be much improved if you add a little, and much less likely to stick. In colder areas fermentations can slow right down over a cold winter and restart again as the weather warms up.

Racking Off

The first racking should be into another clean vessel, trying to leave behind as much yeast as possible and with the minimum of aeration to the cider. This is generally done with a clean plastic syphon tube fixed to a plastic rod so it rests just above the yeast deposit or, on a larger scale, with a suitable pump. Try to rack off with the least amount of disturbance to the sediment, and a cold morning will ensure that the yeast falls to the bottom of the vessel. The transferred cider should be run gently and slowly into the bottom of the new vessel without splashing. Now it is vital to minimise the headspace and to prevent air contact as much as possible. This is why some people add 50 ppm of sulphur dioxide when racking, although the existing carbon dioxide and very limited headspace means this strictly shouldn't be necessary.

Maturation

After racking the air-lock is re-fitted until it is clear that gas evolution has ceased, when the vessel should be topped up with water or cider and tightly closed. A second crop of yeast may be thrown as the cider settles down. The cider may remain in this state for several weeks, before a final racking to a closed container for bulk storage or directly into bottle. Don't let it sit on the lees for more than 2 weeks.

A finished cider will benefit from maturing for at least 2 months (we do a minimum of three months) as its flavour balance stabilises and the harsher notes are smoothed out by slow chemical and biochemical reactions.

Pasteurisation

If you choose to bottle your cider, you can pasteurise at 68 °C for 20 minutes. The cider will be safe to drink for 18 months to 2 years. The pasteurisation process will kill all yeast and bacteria. To pasteurise correctly, you need to ensure the correct temperature and timing and also place capped bottles on their sides to pasteurise inside the neck and cap. Allow to cool naturally.

Cider making Flowchart

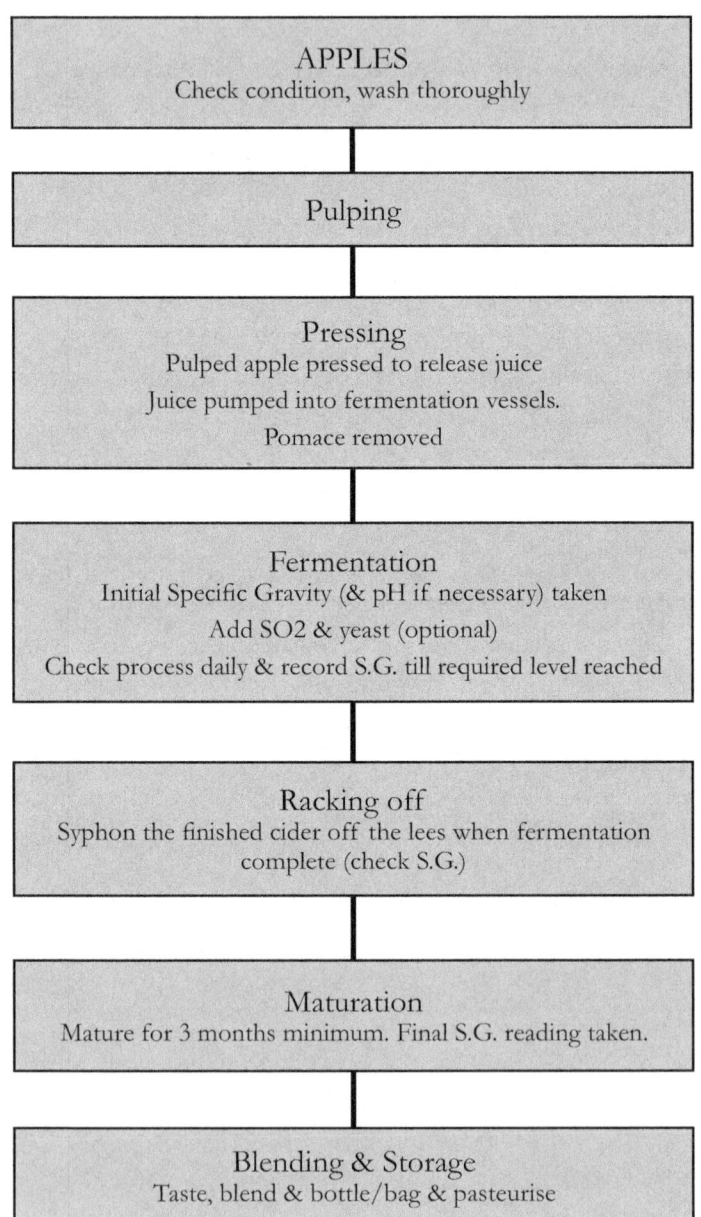

APPLES
Check condition, wash thoroughly

Pulping

Pressing
Pulped apple pressed to release juice
Juice pumped into fermentation vessels.
Pomace removed

Fermentation
Initial Specific Gravity (& pH if necessary) taken
Add SO2 & yeast (optional)
Check process daily & record S.G. till required level reached

Racking off
Syphon the finished cider off the lees when fermentation
complete (check S.G.)

Maturation
Mature for 3 months minimum. Final S.G. reading taken.

Blending & Storage
Taste, blend & bottle/bag & pasteurise

KEEVING

Keeving is a complex artisan method to make a naturally sweetened cider using only cider apples. In the normal cider making method apple juice is fermented out to dry and then sweetened either artificially or with sugar or apple juice. Keeving prevents the yeast from fully fermenting the apple juice and thus some natural sweetness remains when the cider is made. It is a long and slow process and thus not really commercially viable for the big cider producers in the UK. The apples need to be high tannin (mainly bittersweets) apples with low nutrient levels (usually coming from old orchards) and best harvested from late-season trees. The sugar level of the juice needs to be at least S.G. 1.055.

Pulping is undertaken on a cold day, and the pulp stored for 24 hours to allow *cuvage* to take place. This allows the pectin within the cells of the apples to leach out. Then the apples are pressed and the juice collected into fermentation barrels. No yeast is added. A slow fermentation begins to take place and also the formation of a pectin gel. This traps essential nutrients forcing the early stopping of fermentation. The gel combines with the brown foamy cap (the *chapeau brun*) which starts to form and floats to the top of the juice. Some of the pectin and fermented out apple fall to the bottom of the barrel as sediment, and the cider sits neatly between the two.

This cider is carefully syphoned into a clean and sulphited vessel for further fermentation (again no yeast is added) which can take a good few months. It should be racked when reading 1.030 S.G. and then racked one more time when there appears to be no further fermentation visible in the form of bubbles (around S.G. 1.015 or 1.010). Bottling takes place on a cold, dry day to allow for a small amount of bottle conditioning. Traditional champagne bottles, caps and cages are used. In France cider made this way is called *cidre buché*.

For the keen amateur cider maker I can suggest maximising your chances of creating keeved cider by adding the enzyme PME, plus calcium chloride which is available as a French kit sold under the name *'Klercidre'* immediately after pulping your apples (try and source bittersweet apples from an old orchard or this just won't work). The PME and calcium chloride must never be added together but as two separate sequential steps.

You can buy the keeving kit in the UK from Vigo (see the link here) **http://www.vigoltd.com/Catalogue/Chemicals-additives/Keeving-Kit-94426**

WINE

Most winemakers will not have access to a quantity of grapes, and in the UK we normally substitute other fruits, vegetables and flowers to make what are known as 'country' wines. Making wine from your own grapes is pretty straightforward and there is a lot of information on the internet, although some of this information is over simplistic, which is not ideal if you want consistent and drinkable results. A basic wine recipe and method is the basis for many of the recipes in this book, although grapes can be substituted for almost any variety of fruit, vegetables, flowers, spices and even tea. If you wish to make a wine purely from grapes use the recipe below.

Pick 6-7 kg of ripe grapes for every gallon you wish to produce, (or you can buy cartons of grape juice if you wish to cheat a little).

Strip from the stalks, wash, and then crush and extract juice (A potato masher works well).

Check the S.G. and if less than 1.080 add/dissolve some sugar in 5 g increments. Between 1.080 and 1.085 should result in a finished wine of around 11-12 % ABV.

Check the acidity of this juice with a pH strip. You're looking for a level of around 3.1-3.4. You may need to add Citric Acid (or a squeeze of lemon juice) to increase acidity or precipitated chalk to lower it.

Add 1 crushed Campden tablet per gallon of juice and leave for 24 hours. Then add your wine yeast, stir gently and leave to ferment at room temperature in a demijohn filled to the shoulders and fitted with an airlock. Red or rosé wine requires you to leave the skins in for up to 10 days depending on the depth of colour required, before straining them off (paper coffee filters are pretty good to get a really clear wine) and continuing the ferment in a clean demijohn and airlock. Fermentation usually takes 2-6 weeks in a temperature of around 20 °C.

Once fermentation is finished or you've reached the level of sweetness you require (check your hydrometer, and taste), siphon into a sterile demijohn, add 1 crushed Campden tablet per gallon and leave to clear. Mature for 3-6 months, then on a cold day syphon off into sterile bottles, leaving very little headspace.

The process is much the same for any fruit or flower wines you'll make, unless the recipe or method says otherwise. Adding raisins to a floral or insipid fruit or vegetable wine will give it some body.

Keep notes of amounts, when you collected, how long you fermented and how long you matured your wine and remember Cleanliness is King!

RECIPES A-Z

APPLES

Cyser

Cyser is a type of mead that has been fermented with apple juice instead of water, for recipe see Mead.

Mulled Cider

Ingredients:
5 litre cider
2-3 cinnamon sticks
2 star anise
6 cloves
1 orange
5 tbsp honey
1 Vanilla pod
A splash calvados or apple brandy (optional)
Sliced inch of ginger root (optional)

Method:

1. Pour the cider into a pan and heat gently. Cut the orange into quarters and stick the cloves evenly into the skin.

93

Add the orange and clove quarters, the cinnamon, vanilla pod and star anise into the warm cider and allow to infuse for around 20 minutes.

2. When you are ready to serve remove everything from the cider (a slotted spoon is useful), add the honey and the calvados, stir and serve into mugs.

Verjuice

Verjuice is a non-fermented acidic sour juice made from unripe apples or grapes. It was commonly used as a condiment for salads (like we use vinegar today). You'll need around five kg of grapes or apples to make around two or three litres of verjuice. It will keep in a refrigerator for two or three days or you can pasteurise it.

Apple Wine

Ingredients:
3 kg apples
1.5 kg sugar
1 lemon
250 g chopped raisins
1 sachet wine yeast
1 gallon water

Method:
1. Wash and cut up ripe apples into chunks. Boil in a pan with a gallon of water.

2. In the fermentation bucket, put the sugar, and the lemon zest.

3. Strain the apples through a straining bag and dispose of the contents(good animal supplement for sheep, pigs and cattle). Pour the apple liquid into the fermentation bucket, and allow to cool, adding the lemon juice now.

4. Stir and then sprinkle on the yeast, leave lightly covered overnight and then transfer into a demijohn with an airlock. Leave to ferment for around a month, then siphon off into a clean demijohn and add the chopped raisins, and replace the airlock.

5. Allow this to ferment further and when finished siphon off into clean bottles, store in a cool place to mature.

BANANAS

Banana Rum

Ingredients:
2 peeled and sliced ripe bananas
375 ml light or white rum
125 ml water
100 g muscovado or Demerara sugar
Vanilla pod

Method:

1. Peel and slice the bananas and put them and the vanilla pod in a large screwtop jar. Pour in the rum, seal the bottle and shake.

2. Let the mixture settle in a cool, dark place for 3 days. Strain out the liquor – a muslin is useful for the initial straining, and a paper coffee filter for finer filtering.

3. Put the water and sugar into a pan and bring to the boil, stirring to dissolve the sugar, and create a syrup. Leave to cool.

4. Mix the cooled syrup and banana rum liquor together in a bottle, seal and then shake to mix. Store in a fridge for up to one month.

BAY

Bay and Rosemary Ale

This makes a sweetish fruity ale with a bit of a kick. Beware that the bay may release the inhibitions of your guests, but this is a nice easy-to-drink beer for an informal party or a wassail. If you like your ale with a bit more bitterness you can add a pinch of hops, but the bay does add a nice kick.

Ingredients:
200 g malt extract (do NOT try molasses for this recipe)
2 fresh rosemary sprigs
4 fresh bay leaves
100 g sugar
honey or golden syrup for priming
Ale yeast
1 gallon (boiled then cooled) water.

Method:

1. Bring the water to the boil and add the malt extract, then strip the rosemary and bay leaves into the water and keep boiling for 30 minutes (you can add the hops now if you fancy them).

2. Remove from heat, add the sugar and stir until full dissolved.

3. Strain the liquid into a fermentation bin. Pour over 6 litres of cold water and allow to cool. An initial S.G. reading should be about 1030. Sprinkle on the

yeast and loosely cover. Allow to ferment for up to 14 days.

4. Siphon into bottles using one tsp golden syrup or honey or 1/2 level tsp sugar as primer per bottle. Leave for at least a week before drinking.

BEECH

Beech Leaf Noyau

This sweet and nutty gin-based liqueur was made and drunk by forest workers in England, although *noyau* is a French word for the stones of peaches. As this is uncommon in the UK, I suggest making enough to fill a litre glass bottle in order to try it rather than make a demijohn full.

Ingredients:
750 ml bottle of supermarket gin.
350 ml brandy
200 g sugar
Large handful of clean very young beech leaves

Method:

1. Pack the beech leaves into a jar and pour in the gin, covering the leaves fully. Put the lids on, and store in a cool dark place for 3 weeks.

2. Strain the gin into a large jug or pan, add the brandy and the sugar. Stir to fully dissolve the sugar, bottle (Makes about a litre). Ready to drink in one month. Serve chilled.

BEETROOT

Beetroot Wine

Beetroot makes an earthy sweetish wine. This is one of those love it or hate it wines. If you like beetroot as a vegetable, you'll probably love this. It can be improved by adding rhubarb when making or when topping up the demijohn after the final racking. Warning – this will stain everything whilst you are making it and as a finished wine.

Ingredients:
1.5 kg Beetroot (trimmed, washed and green parts removed)
4 litres water
1.2 kg Sugar
100 g Raisins (roughly chopped)
1 cup of apple juice/a couple of slices of green apple or some rhubarb chunks
Zest and juice of 2 limes
Small cup of black tea
1 tsp Pectolase
1 tsp Yeast Nutrient
1 Campden Tablet
1 Sachet Yeast

Method:

1. Wash the beets (small ones are sweeter) and twist off the leaves. Slice thickly and put into a large pot and just cover the pieces with water. Bring to a simmer and cook for 40 minutes or until tender. Turn off heat.

2. Lift the pieces from the liquid and set aside. Add the sugar to the pan of liquid and stir to dissolve.

3. Take the raisins and a few slices of the cooked beetroot and place in a straining bag with the lime zest (and apple or rhubarb), and place this bag inside a fermentation vessel securing the bag at the top of the bucket.

4. Pour on the sugar solution. Add the lime juice, apple juice if using this, the tea, yeast nutrient and 1 crushed Campden tablet. Cover and leave for 24 hours.

5. Add the pectolase and stir to mix. Take a hydrometer reading if you wish. Sprinkle on the yeast. Leave to ferment for a week, then remove the bag with the raisins and beetroot. Cover and refit the airlock and leave for a further 2 days.

6. Transfer the liquid into a demijohn and top the level up to the shoulders with water if necessary.

7. Ferment out till finished, which may take a month, then rack off into a clean demijohn and take a second hydrometer reading. Top up the demijohn with water (or finished rhubarb wine) to the neck and fit a clean airlock. Leave to settle and clear in a cool place for 3 months. Bottle and store for at least a year before drinking.

Beetroot and Gooseberry Wine

This recipe just smooths out the earthy taste of the beetroot a little with some gooseberry notes. A tart dry wine which can be back-sweetened after finishing.

Ingredients:
Juice from a mix of gooseberries and cooked beetroot.
250 ml white grape concentrate or a handful of chopped raisins
750 g sugar
1 tsp nutrient
1 tsp pectolase
1 sachet wine yeast
Campden tablet
Water to 1 gallon

Method:

1. For best results use fully ripe gooseberries. Popping them into a freezer bag and overnight in the freezer, then allowing them to defrost the next day releases the juice more easily.

2. Wash and cook the beetroot, drain and allow to cool. Slice. Put the gooseberry juice & fruit and the cooked beetroot in a nylon straining bag inside a fermentation bucket and pour over the boiled water. Add and stir the sugar until dissolved. Allow to stand until cool.

3. Drain the straining bag over the bucket to collect the juice, then pour this liquid into the demijohn.

Add the white grape concentrate, nutrient and stir to mix, then sprinkle on the yeast.

4. Ferment in the usual way under an airlock, rack, add the pectolase and rack the wine before bottling. Mature for at least a year.

BERRIES (Rowan, Hawthorn)

Hawthorn Wine

Ingredients:
2 kg berries
1 lemon
2 oranges
1 kg sugar
1 sachet wine yeast
4.5 litres water
Pectolase

Method:

1. Put the clean berries, the zest and juice of the lemon and the oranges in a large plastic bucket and pour over 4.5 litres of boiling water.

2. Leave this to soak, covered for a week.

3. Strain the juice through a straining bag into a clean bucket. Add the sugar and stir to dissolve. Sprinkle on the yeast and cover lightly, and leave for 24 hours.

4. Transfer into a demijohn, fit an airlock and ferment to finish. If the wine seems a little cloudy, add the pectolase, and after a few days rack off into a clean demijohn, top up to the neck with water and allow the wine to settle.

5. Syphon off into bottles and store for at least a year before drinking.

Rowan or Hawthorn Schnapps

Ingredients:
A few handfuls of fresh, ripe rowan berries
Vodka
1 tbsp of sugar

Method:

1. Rinse the berries thoroughly.

2. If you're harvesting before the first frost, you could put them in the freezer for a couple of weeks – it may make them sweeter, and will soften the skin.

3. Fill two thirds of the jar with the berries and then add the vodka and sugar.

4. Screw the lid on tightly and shake the jar. Store it in a cool, dark place for 4 weeks, shaking the jar every few days. Strain and bottle, and it's ready to drink in 8 weeks.

BRAMBLES OR BLACKBERRIES

Bramble Whisky Liqueur

Ingredients:
A handful of blackberries (or raspberries)
Sugar
Whisky

Method:

1. Place your blackberries or raspberries in a sealable jar and add about a tenth of their volume in granulated sugar.

2. Top up with whisky to cover the fruit. A supermarket own-brand will be perfect. The next part is the hardest. You have to leave it for six months somewhere cool and dark, giving it only the occasional shake.

3. After six months have passed, strain the liquid out of the blackberries. A muslin cloth or paper coffee filter will catch all of the solids.

4. Taste and sweeten a little if necessary at this point. Bottle your liqueur.

Blackberry Wine

Ingredients:
1.5 kg blackberries or brambles
1.5 kg sugar
1 lemon
1 sachet red wine yeast
1 gallon water
Pectolase (optional)

Method:

1. Pick over and wash the blackberries and place in a fermentation bucket with the zest of the lemon.

2. Pour on the gallon of boiled water, cover lightly and let it stand for 3 days, stirring daily.

3. Strain the mix into a clean fermentation bucket, add the sugar, (pectolase if you wish a really clear wine) and the lemon juice and stir.

4. Sprinkle on the yeast, cover loosely and leave for a day.

5. Pour into a demijohn and insert an airlock. When finished fermenting siphon off into a clean demijohn, top up with water if necessary to neck, fit a new airlock and wait to see if there is any further fermentation.

6. Syphon off into bottles and mature.

CARROTS

Carrot 'Whisky' (really a wine)

This makes a slightly hazy wine rather than a whisky (the name carrot 'whisky' comes from the colour at finish). You need to mature for at least 18 months before bottling to drive off the harsh ethanol taste.

Ingredients:
3.25 kg topped carrots, washed but not peeled, grated (I used a food processor).
5 litres water
550 g malted wheat grains or cracked wheat (burgul)
Zest and juice of 3 oranges
Zest & juice of 2 lemons
1.5 kg sugar
Pectolase
1.5 tsp yeast nutrient
1 sachet wine yeast.

Method:

1. Wash and top the carrots and blitz in the food processor.

2. In a large pan bring to the boil with enough water to fully cover them. Simmer for 30 minutes.

3. Add the zest and juice of the oranges and lemons to a straining bag, together with the wheat, and fix this into a fermentation bucket.

4. Drain the liquor from the carrots into the fermentation bucket. Add the sugar and stir gently to dissolve, and then add the rest of the water. Allow to cool.

5. Add the pectolase and leave covered overnight. The next day add the yeast nutrient and sprinkle on the yeast. Cover loosely with the lid and leave in a warmish place, stirring gently daily for a week.

6. When the fermentation seems to have slowed down a little remove the straining bag and allow it to drip the liquor into the bucket for a few hours. Transfer this liquid into a demijohn and fit an airlock. Allow to ferment out for a couple of months. If it throws a lot of lees after a month or so, rack off into a clean demijohn and top up with water to the shoulders. Fit a clean airlock.

7. When fermentation seems to have completely finished, rack off and bottle and allow to mature for at least 18 months. The colour will fade to a whisky colour and the harshness of the flavour will mellow.

Carrot Wine

Carrots make a delicious wine, and even if you don't yet grow your own, they are dirt-cheap to buy. Okay, let's be honest and accept that it's not going to taste like a good Chardonnay, but you might be surprised by how nice it actually is. Smaller carrots make for a sweeter wine, but big woody ones aren't too bad. The recipe below makes a gallon.

Ingredients:
2.5 kg carrots
4 litres water
900 g sugar
400 g honey
250 g chopped sultanas or raisins
½ tsp Pectolase
1 tsp Yeast nutrient
1 sachet white wine yeast
1 Campden tablet

Method:

1. Scrub the carrots and put in a food processor to chop finely, then boil in a large pot with 2 litres of water.

2. Reduce heat and add the sugar and honey. Simmer this mix for 10 minutes, then turn off the heat and add the sultanas. Allow to cool.

3. Prepare a large fermenting bucket and place inside this a fine straining bag or muslin. Pour the contents of the pan into the bucket, securing the straining bag or muslin, but leave in the bucket.

4. Add the remaining water (2 litres) and add the crushed Campden tablet. Cover and allow to sit for 24 hours.

5. Add the Pectolase and mix in the liquid, then sprinkle the yeast sachet on top of the liquid. Cover the bucket with the lid and add an airlock. Fermentation should continue for around 7 days, after this time, lift the straining bag and discard the contents. Allow the wine to sit overnight in a cool place.

6. Rack the wine into a demijohn, filling to the shoulders, fit an airlock and allow the wine to finish fermentation. This may take 3-6 months.

7. If the wine throws more sediment, simply siphon into a clean demijohn again, and top up with water to the neck. After the final racking, taste, sweeten if necessary with a non-fermenting sweetener like Splenda and bottle. Mature for 6 months.

CHERRIES

Cherry Brandy

Ingredients:
450 g cherries
85 g sugar
150 ml brandy

Method:

1. Wash cherries and place into a large screwtop jar. Add the sugar and gently pour in brandy.

2. Gently shake or stir once a day for two weeks or until the sugar has all dissolved. Store in dark place for 3 months.

3. Filter into bottles. Reuse the spent cherries by half-dipping in dark chocolate. They can be frozen.

COURGETTES

Courgette Wine

Yes, there really is a wine you can make with a courgette glut (thank God). If you are planting this in the garden, one plant is ample. They need a bit of room and a free-draining, but moist soil. The bananas are necessary to add a bit of body, and the ginger gives it a kick.

Ingredients:
3 kg fresh courgette or summer squash
1 kg sugar
2 over-ripe bananas
500 g sultanas or raisins
4 litres of water
1 tsp acid blend
A 3 cm length of ginger root, bashed/bruised
1 tsp yeast nutrient
1 Campden tablet
1 sachet white wine yeast

Method:

1. Wash the courgettes and blitz in a food processor. Put in a freezer bag and freeze overnight.

2. Allow the courgettes to thaw out the next morning whilst you boil 500 g of water in a pan.

3. Turn off the heat, add the sugar and stir until dissolved. Add the sultanas or raisins to the sugar water and leave to rest.

4. Mash the bananas in a fermentation bucket and add all the other ingredients including the thawed courgettes *except* the yeast. Stir and allow to cool. Add the crushed Campden tablet, and cover and leave overnight.

5. Sprinkle on the yeast, and cover loosely for 4 days, allowing fermentation to begin. Place the bucket in a warmish area.

6. Strain the liquor through a straining bag and pour into a demijohn and fit an airlock. Fermentation will slow and may take up to three months to ferment out. Check the airlock is filled with water. You do NOT want oxygen spoiling your wine now.

7. After one month do the first racking off into a clean demijohn, leaving the lees behind. Top up with water to the neck, fit a new clean airlock and leave to ferment out. After two more months move the demijohn into a warm area and check the next day if there are still any fermentation bubbles in the airlock. If not you can bottle; if there is, return it to finish fermenting in a cool, dark place.

8. Mature the wine for at least 3 months before drinking.

CURRANTS

Cassis

Way back in the 1800's in France a Frenchman started to make crème de cassis commercially, although it was probably a well known home-made treat. You can make your own version with this recipe. I've tried making this with gin, but it's much nicer with a vodka base.

Ingredients:
350 g blackcurrants
100 g Sugar (demerara is quite nice)
500 ml vodka

Method:

1. Wash and destalk the blackcurrants and place them in a large jar with a tight fitting lid. Pour in the sugar and then the vodka. Tighten the lid then shake thoroughly. Store in a cool dark place, shaking twice a week. Leave for 3-6 months, strain, but do not squeeze the fruits (a paper coffee filter will give a cleaner liqueur) and bottle.

Redcurrant Wine

Ingredients:
2 kg redcurrants
1.5 kg sugar
Pectolase
1 sachet red wine yeast

Method:

1. Wash and de-stalk the currants and place in a large food-grade bucket with lid.

2. Mash the fruit with a potato masher and pour on 2.5 litres of boiling water.

3. Cover and leave to soak for 2 days.

4. Strain through a straining bag into a clean bucket, squeezing as much juice as possible.

5. Add the sugar to this liquid and stir to dissolve. Add the pectolase. Pour into a demijohn and top up to the shoulders with water if necessary, and sprinkle on the yeast.

6. Allow to ferment in a darkish place and when finished fermenting rack off into bottles.

DAMSONS

Damson Rum (You can substitute sloes or a mix of the two)

This is an infusion, just like a flavoured gin or sloe gin or vodka. Cut your clean damsons into two (if using sloes, pop them in a freezer bag and freeze over night) and put in the bottom of a screwtop jar, pour over the rum and tighten the lid. Shake and then leave the jar in a cool, dark place for two weeks, and then strain off the liquor through a coffee filter into a clean bottle. The damsons can be used in a fruit pie, but the sloes best discarded.

DANDELION

This is one of my favourite things to ferment. They are readily available, free and easy to use, and can be used to make beer and wine.

Dandelion Beer

I actually really like this recipe, and so does everyone who's tried it. The preparation makes it look a little foul, but as a cold late spring drink it's refreshing and moreish.

Ingredients:
225 g whole, young dandelion plants
450 g Demerara sugar
10 g root ginger
1 large lemon
25 g cream of tartar
1 sachet white wine yeast
1 gallon water

Method:

1. First gather your dandelions. Dig up the whole plant, keeping the roots intact. Big fat roots are best, so choose the biggest plants. Wash thoroughly to eliminate all soil, and remove any threadlike roots. Peel and chop the ginger, and peel the lemon.

2. Put the entire plants into a large saucepan with the water, ginger and the peel of the lemon. Bring to the boil and simmer for 10 minutes.

3. Strain through a bag or muslin into the fermentation bucket. Add the sugar and the cream

of tartar and mix thoroughly. Allow to cool to room temperature.

4. Sprinkle on the yeast and lightly fit the lid. Leave for 3 days in room temperature to ferment, then siphon off into screw top plastic bottles. Store in a COLD place and watch the pressure; you may have to release the gas carefully once or twice daily. Ready to drink in one week.

Dandelion and Burdock Beer

If you are of a certain age; like me, you may remember your granny making this. As no-one grows Burdock in their garden you will have to forage for this. Ensure you wait until the leaves have fully unfurled to enable a correct ID, as the burdock can be mistaken for the fatal Hemlock Water Dropwort; and you'll need a spade as the roots go deep. This recipe brews a mildly alcoholic bitter-sweet beer.

Ingredients:
150 g Burdock roots
50 g dandelion roots
1 tsp dried carragheen
500 g sugar
5 g of peeled bashed ginger root
1 star anise (optional)
2 tbsp. black treacle
Juice of 1 lemon
1 sachet ale yeast
4.5 litres water

Method:

1. Thoroughly wash the burdock and dandelion roots and chuck into a food processor to finely chop. Peel and bash the piece of root ginger.

2. Put all these in a large pan and pour on 2.5 litres of boiling water, and add the carragheen and star anise if using. Boil for half an hour.

3. Take off the heat, add 2 litres of cold water, the sugar, treacle and lemon juice and stir until the

sugar has dissolved. Strain the liquid into a clean fermenting bucket, cover loosely and leave to cool.

4. When your brew reaches room temperature, add the yeast. Cover and leave to ferment for a week, then bottle in plastic screw top bottles, releasing the pressure slightly if the bottle feels very hard and rigid.

5. After a week this pressure build up should have slowed right down and you can store the beer in a cold place. Just keep an eye on the pressure in the bottles.

Dandelion Wine

Ingredients:
3 litres of dandelion flowers
1 kg sugar
1 orange
2 lemons
1 tin white grape concentrate
1 cup strong black tea
1 sachet of white wine yeast
1 gallon water

Method:

1. Pick the flowers on a warm sunny day, remove as much green from them and then pack them and the zest of the orange and lemons into a straining bag, and secure this in a fermentation bucket.

2. Boil the water in a pan and dissolve in the sugar. Add this to the bucket along with the juice of the orange and lemons, the cup of black tea and the grape concentrate. Stir and leave to cool.

3. Sprinkle on the yeast, cover loosely and allow to ferment for 3 days and then remove the bag and discard the contents.

4. Transfer the liquid into a demijohn and fit an airlock. Leave to fully ferment and then rack into lean bottles. Store in a cool place and should be ready to drink by Christmas.

ELDER

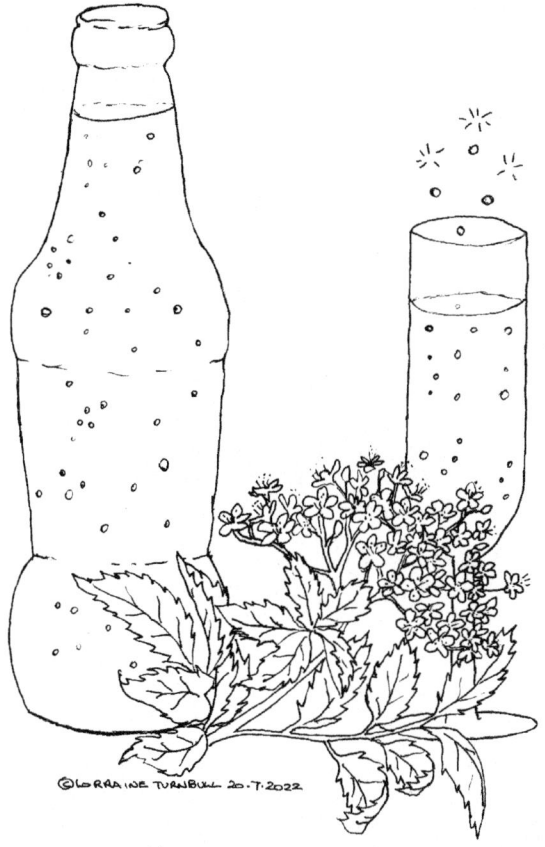

Remember in the foraging section, I said that ID was crucial if you are not to poison yourself or anyone else? Well, making the correct ID on a plant like elder is crucial. True elder (*Sambucus nigra*) is safe to use, but its cousin *Sambucus ebulus*, the Dwarf Elder or Danewort has toxic berries, and must not be used. Sadly, they look very similar and grow in the same places, so make sure. When picking elderberries, try and pick when there has been a couple of dry days, as too much moisture in the berries makes them melt into inky puddles.

Elderberry Port

This recipe makes about 5 bottles of strong port-like wine. Don't top up the demijohn with water – use brandy or port instead.

Ingredients:
4 pints of elderberries or elderberry/ blackberry mix (approx 2.3 litres)
1.5 kg sugar
100 g raisins
1 banana
Juice of one orange
1 gallon of water
1 sachet burgundy or general red wine yeast
Yeast nutrient

Method:

1. Strip the elderberries from the stalks, a fork makes this an easier task. If you immerse the berries in cold water for a few minutes, any hidden insects will float off.

2. Place into a large pan and crush with a potato masher. Add the water and bring to the boil, then simmer for 15 minutes, stirring to prevent sticking.

3. Strain through a sieve into a fermenting bin whilst still hot and add the sugar and raisins, stirring well to dissolve the sugar. Allow to cool.

4. Take an initial S.G. reading, looking for 1.09 (add sugar if necessary to reach this, stirring to mix). Add the wine yeast and nutrient. Cover loosely with

the lid and leave to ferment for three weeks, then siphon into a demijohn leaving the fruit and lees behind. Fit with a clean airlock.

5. Ferment until finished, and rack off into a clean demijohn. Take a final S.G. reading at this time and add 1 crushed Campden tablet to stop secondary fermentation. Instead of topping up with water, top up to the neck with brandy or port and mature for two to three years.

Elderberry Wine

As autumn approaches and you find yourself in front of the fire, it's the perfect time to enjoy a glass of elderberry wine. Patience is required for this recipe. Like red wine grapes, elderberries are high in tannin and so the wine needs time to mature.

If you want this wine to be more like port, then ensure the berries are fully ripe and dry, and use 1.5 kg sugar to 2.5 kg of berries, and add 250 g chopped raisins when it goes into the demijohn.

Ingredients:
1.5 kg elderberries
Juice of 1 lemon
2 Campden tablets
1 kg, 1.2 kg, or 1.5 kg sugar (use the lowest amount of sugar for a dry wine, next for a medium dry wine and the higher sugar for a medium sweet)
1 sachet of red wine yeast
Yeast nutrient
3.5 litres water
Gloves - particularly when handling the berries as they can die your skin purple!

Method:

1. Strip the elderberries from the stalks and wash well, then put into a fermenting bin and crush, either using gloved hands or a potato masher.

2. Pour on 2 litres of water and add 1 crushed Campden tablet (optional), and stir to mix.

3. Boil half of the sugar in 1 litre of water for 2 or 3 minutes and, when cool, mix into the pulp. Add the lemon juice now.

4. Add the yeast and nutrient and cover and allow to ferment for 5 days, stirring daily. The initial fermentation may be quite lively, so choose your fermentation place with care. This is not a wine to house in the airing cupboard. Strain firmly and return the liquor to a clean fermenting bin.

5. Boil the rest of the sugar in 500 ml of water for 2 or 3 minutes and when cool, add to the liquor.

6. Cover again and leave for another 3 or 4 days.

7. Syphon into a demijohn. Try and leave as much of the sediment behind as possible.

8. Fill up the jar with cooled boiled water to where the neck begins and fit an airlock and leave until fermentation has fully finished, which could take a couple of months.

9. Rack your wine (which means to move your wine into a fresh container) adding a crushed Campden tablet. Syphon into bottles, label and date them. Ready to drink in a year, but better at two years.

Elderflower Fizz

Ingredients:
10 large heads of elderflowers
Juice and rind of 1 lemon
2 tbsp white vinegar
1 sachet champagne yeast (optional)
500 g sugar
4 litres water

Method:

1. Put everything in a large fermentation bucket except the yeast & water. Warm the water and add it, stirring to dissolve the sugar.

2. Now, the elderflowers will have a small amount of natural yeasts on them if you collect on a warm day when the flowers are fully ripe, so you can omit the sachet of yeast if you wish. Sprinkle on the yeast, cover loosely with a clean muslin and leave for 4 days, stirring occasionally.

3. Strain through a straining bag or muslin into plastic screwtop bottles. Do not tighten the lids fully, but store somewhere cool and check the pressure daily. This is ready to drink in 3 days, but definitely drink within 3 months. If you wish to make this throughout the year, you can collect and freeze the elderflower heads.

Elderflower Wine (strong)

Ingredients:
4 large elderflower heads
200 ml white wine concentrate (or
200 g of raisins)
1 kg sugar
5 g citric acid or the juice of a big
lemon
1 cup strong black tea
1 sachet white wine yeast

Method:

1. Strip the flowers off the stalks into a bucket (the more green you remove the better it will taste).

2. Pour 5 litres of boiling water on top of this, and stir a couple of times over 24 hours. Strain through a straining bag or muslin into a clean bucket.

3. In a pan dissolve the sugar in 500 ml of boiling water, and add this to the elderflower liquid in the bucket, and add the wine concentrate, the black tea and the citric acid. If you substitute raisins, these will float during fermentation, so when you rack off you'll have to push the siphon tube under the raisin layer.

4. Pour into a demijohn filling to the shoulders, sprinkle on the yeast and fit an airlock. Leave in a warmish place to ferment out.

5. When fermentation has stopped, siphon off the wine into a clean demijohn, leaving the lees behind.

Top up the new demijohn to the neck with a little water and fit a new airlock.

6. Leave to mature for about 6 months, and rack into bottles, filling to 5 cm from the top of the bottle.

Elderflower Wine (mildly alcoholic)

Ingredients:
6 elderflower heads
2 lemons
750 g sugar
4 litres of water

Method:

1. Roll the lemons on a hard surface to break up the peel and the internal juice. Slice each in half, juice them and save the juice and put the spent lemon halves in the fermentation bucket with the elderflower florets. Boil the water and pour on.

2. Leave the liquid to soak for 24 hours, covering the bucket loosely.

3. Strain through a straining bag into a clean bucket, add the sugar and the lemon juice, stirring until dissolved.

4. Pour into two 2 litre plastic screwtop bottles, leaving the lids not quite tightly shut. Leave in a cool place for two weeks, then when fermentation seems to have finished screw the lids tight and it is ready to drink in 2 months.

FIGS

Fig Wine

This makes an unusual sweetish wine. When picking ripe figs from a tree wear gloves as the sticky white sap is a skin irritant. And I urge you to drink this wine in small doses. If you are fortunate to be of the younger generation you will probably be unaware of the laxative qualities of fresh, dried or fermented figs. So, unless you have a particular *pressing need* to drink more than a small amount of this wine for health reasons, or you wish to give it as a revenge gift to someone you dislike, it's a wine to drink cautiously.

Ingredients:
2.0 kg figs
3 litres water
800 g sugar
3 tsp acid blend
1 crushed Campden tablet
1 tsp. yeast nutrient
1 sachet red wine yeast

Method:

1. Cut off stalks and chop figs, and place in large, straining bag, tie the top, and put in your fermentation bucket. Warm the water and add this and all other ingredients except yeast.

2. Take an initial S.G. reading (should be 1.085 to 1.095; if not, add a little more sugar in 10 g increments, stir and take further readings). Cover with sanitized cloth and leave overnight.

3. Sprinkle on the yeast and stir twice daily for 3-5 days until the liquor produces a S.G. reading of 1.040 and then remove the bag, allowing it to drip over the bin to catch the juice. Don't press the fruit or you will make a cloudy wine.

4. Syphon the liquor in the bucket into a demijohn, fit an airlock and ferment for a round three weeks until the liquid gives an S.G. reading of around 1.000 or lower. Rack off into a clean demijohn, topping the level with water to the neck, re-fitting an airlock and leaving to further rack two months later. Bottle and leave to mature for 6 months.

GINGER

Now ginger is a strange ingredient. Back in the olden days, when we all had open fires, no central heating and everything was damp; a lot of home-brew recipes traditionally added ginger (as a root or as ground ginger) to give alcohol a bit of a kick, but tastes have changed and modern recipes are less likely to include it (and far less sugar too). If you find very old booze recipes and some contain ginger, you could omit it altogether or at the very least reduce the amount.

The recipe for alcoholic ginger beer is in the Quick Booze section towards the rear of the book.

Ginger Infused Gin

Ingredients:
750ml bottle gin
3cm x 7cm piece of peeled and sliced
or grated fresh ginger root
5 crushed cardamom seeds

Method:

1. Combine all the ingredients in a large screwtop jar, seal the lid and shake. Leave in a cool, dark place to infuse for 7 days.

2. Strain out the ginger and cardamom and bottle.

GORSE

Gorse or Broom Wine

Ingredients:
2 litres gorse or broom flowers
1 kg sugar
2 lemons (preferably unwaxed)
2 oranges
4.5 litres water
1 sachet white wine yeast

Method:

1. Pick fresh fully-opened flowers. Due to the prickly nature of gorse, this is easiest on a sunny day with no breeze. When you get your flowers home shake to remove any beasties.

2. Put the flowers in a large pan and cover with the water, bringing up to a simmer, then turn off the heat. Add the sugar and stir to dissolve. Allow to cool to room temperature.

3. Pour into a fermentation bucket, add the zest and juice of the lemons and oranges, and then add the yeast. Stir and cover loosely with the lid. Leave for 3-4 days stirring once a day.

4. Strain off the liquor into a demijohn, fit an airlock and allow to ferment for at least four months, topping up the airlock if required.

5. Rack off into a clean demijohn, filling with water to the neck and refitting the airlock. Leave to stand for

1 more month. Bottle when clear or leave another month to clear further. Mature for 6-9 months. Back-sweeten if necessary with a non fermentable sweetener.

HEATHER

Dried heather tops (the flowering tips) are quite bitter, almost like dried hops, so if you can find heather you can substitute it for hops.

Heather Ale or Fraoch

Ingredients:
25 g heather tops (the flowering tips in full bloom)
1 tea bag
12 g yarrow or camomile flowers
6 g dried hops (optional)
400 g honey
260 g malt extract
100 g crushed crystal malt grains
1 tsp dried carragheen (Irish moss)
1 sachet ale yeast (Kviek is perfect)
5 litres water

Method:

1. Steep the crystal malt grains in 2 litres of water at 65 °C, cover and leave for 30 to 40 minutes. Strain the grains from the 'wort' or liquor, and dispose of the grains (sheep and cattle love this as a supplement) and pour the wort liquid into a big pan. Drop in the tea bag to steep for 5 minutes, then dispose of this.

2. Add 1 litre of hot water to the wort and bring to the boil. Add the malt extract, yarrow, heather and hops and boil for one hour. After one hour add the carragheen and the honey and boil for another 30

minutes, then leave to rest for further 30 minutes.

3. Strain through a straining bag into a large fermenting bin. Top up to 5 litres by pouring some cold water through the bag and its contents. Cover and leave to cool to room temperature. Stir with a spoon until there is a bit of a froth on the top. Allow to settle then sprinkle on the yeast. Cover.

4. After 24 to 36 hours a cauliflower-like head or cap will have form on the surface. Skim this off and allow the beer to continue fermenting until the S.G. has dropped to 1.010.

5. Siphon into plastic bottles or into a 25 litre plastic pressure barrel. Check every day or so to make sure the pressure has not reached explosive potential, releasing the pressure if necessary. The beer will continue to ferment in the bottle (this is called bottle conditioning), creating more alcohol, reducing the sugar and adding fizz. It is ready to drink after one to two weeks.

HONEY

Mead

Mead is made by fermenting honey with water and is pretty alcoholic at between 12-15%. Metheglyn is mead flavoured with herbs or spices, and melomel is mead flavoured with fruit. Hydromel is a lower alcohol mead. The recipe below will produce a session-type mead of around 4%. When making mead choose the most flavoursome honey you can; heather honey and orange honey are particularly good.

> **Ingredients:**
> 150 g of honey per litre of water. If you want to make 5 litres of mead use 750g of honey.
> 1 sachet ale or wine yeast
> 5 g citric acid
> 1 Campden tablet.

Method:

1. Add 5 litres of boiled water to the fermentation bucket. Add the yeast nutrient and citric acid and stir. Add the honey to the bucket and stir. Leave to cool to warm, and then decant using a funnel into a demijohn. Take temperature and when it has dropped to 20 °C sprinkle on the yeast.

2. Add the airlock and leave in place in a constant room temperature for 2-3 weeks.

3. Keep a watch on the fermentation, and as the bubbles start to slow down. When bubbling has completely stopped your mead is ready to rack off.

4. Place the demijohn in a cold place overnight (or in a fridge) to encourage the yeast to drop out of solution, and then rack the finished mead off the sediment and into a clean demijohn.

5. The mead will have fermented out to dry and may need to be back-sweetened. You can do this with more honey, but will need to stop the fermentation by adding 1 crushed Campden tablet to the demijohn, and leave to condition in a cool, dark place for 2-6 weeks. Bottle and cap it for storage.

Lavender Mead

Ingredients:
1.5 kg honey
125 ml dried lavender flowers
1.5 litres water
1 sachet mead or ale yeast
1 tsp nutrient

Method:

1. Pour the honey into a demijohn, add the lavender flowers and heat the 1.5 litres of water till very hot and pour on top. Add the yeast nutrient and top up with cold water to the shoulders.

2. Sprinkle on the yeast and fit an airlock. Leave to ferment for 2 months.

3. Rack the mead into a clean demijohn, and top this up to the neck with water. Fit a clean airlock.

4. Rack again after 6 months. Top up again to the neck with water if necessary. Leave for a few days to settle then bottle and store for 1 year before drinking.

Cyser

Cyser is a type of mead where the honey is fermented with apple juice instead of water. Use the mead recipe and just replace the water with apple juice.

METHEGLYN

Metheglyn is simply mead that is flavoured with herbs and or spices. It was once used as a medical pick-me-up, especially in Wales. Once you've made your base mead, simply add thyme or lavender or other flavourings, or add them when you are fermenting the mead for a stronger taste.

Heather Metheglyn (an old recipe from my Highland granny)

1. Take the heather bells in full bloom, about 25 g of them, wash in cold water and place in a pot, cover with 5 litres of water and boil for an hour.

2. Strain into a clean pan, and add 5 g of ground ginger and 500g of honey. Boil again for another twenty minutes.

3. Strain off again and cool. Pour into a demijohn or fermentation bucket. Add 1 sachet of ale yeast. Cover with a cloth and leave to ferment for 2 days.

4. Syphon the liquid off from under the head, and avoiding the lees and bottle.

5. Store in a cold place.

6. Ready to drink in 1 week, but much better after 6 months.

MELOMEL

Blackberry Melomel

Blackberries are a good fruit to make a melomel with as they retain the dark colour and tannin of the original fruit which balances the taste of the mead. This recipe makes a gallon of melomel, and the ABV at finish should be around 12-13%.

Ingredients:
1.3 kg ripe blackberries
1.4 kg honey
3 litres boiled and cooled water
Zest of half a lemon
1 sachet red wine yeast
1 Campden tablet
1 tsp Yeast nutrient
1 tsp pectolase

Method:

1. Place the cleaned blackberries and the lemon zest into a straining bag and secure inside the fermentation bucket. Mash with potato masher.

2. Add the honey to the bucket and add the water to make up to 4.5 litres. Stir gently to incorporate the honey, and then add the nutrient and crushed Campden tablet. Take an initial S.G. reading. Cover with the lid and leave overnight.

3. The next day add the pectolase enzyme and stir. Then sprinkle the yeast onto the surface. Cover and fit an airlock to the lid and allow fermentation to begin.

4. Ferment for a week, then remove the straining bag allowing the liquid to drain back into the bucket.

5. Rack off into a clean demijohn leaving a small headspace and secure with airlock.

6. Leave for two months, then rack off again into clean demijohn and allow to condition for four to six months. At this racking take your final S.G. reading.

7. At the end of the six months taste your melomel. If you need to sweeten it further you can use a non-fermenting sweetener. Bottle to within 2 cm of the top of the bottle.

HYDROMEL

Hydromel has a much lower ABV than a normal mead, usually around 4-7 %ABV. So how is this lower ABV achieved? Simply by increasing the quantity of water during the fermentation process. It has a more subtle honey flavour and can ferment out to remove any residual sweetness, leaving it very dry to taste. An artificial sweetener will add a touch of sweetness whilst allowing carbonation. The lack of flavour can be counteracted by adding fruity, floral or spicy notes. The recipe below is for a mixed berry hydromel; finishing around 5 % ABV, but you can use other fruits such as peaches, flowers such as honeysuckle, or a vanilla pod.

Ingredients:
1.3 kg honey
1.2 kg mixed berries
10 litres of water
1-2 tsp acid additive
¼ tsp tannin
1 tsp Pectolase
2 tsp yeast nutrient
1 sachet ale yeast (Kviek is a good one for a this recipe)
Artificial non-fermenting sweetener to back sweeten.

Method:

1. Fit a straining bag around the mouth of your fermentation bucket, and place the prepared fruit inside this straining bag. Mash the fruit if required and secure the straining bag.

2. Boil 4 litres of water and pour into the bucket. Add the honey and stir carefully to incorporate. Add enough cold water to bring the liquid up to 10 litres, then stir in the tannin, acid, yeast nutrient and pectolase. Take an initial S.G. reading.

3. Check the temperature, which needs to be within the range indicated on the packet, but around 20 °C. Sprinkle the yeast onto the surface and fit the lid with an airlock.

4. Ferment for a week and then carefully remove the straining bag and contents. The fruit can be reused for trifle or eaten with ice-cream.

5. Refit the lid and airlock and continue to ferment for another week or so. Take a hydrometer reading to check that fermentation has ceased.

6. If you wish to back sweeten, then now is the time, before you prepare to bottle. A very small amount of Erythritol (15 g per litre) or Splenda will do.

7. Satisfied with the taste you now need to bottle condition to make the finished hydromel sparkling. Prime the individual bottles with sugar (3 g per 500 ml bottle will lightly carbonate), fill with hydromel and cap immediately. It will be ready to drink in three weeks.

LAVENDER

Lavender Wine

Ingredients:
1 to 1½ pints lavender flowers
750 g sugar
250 ml white grape juice concentrate
1 tsp acid blend
4 litres water
1 tsp yeast nutrient
1 sachet white wine yeast

Method:

1. Boil 2 litres of water in a large pan, turn off the heat, add the sugar and stir to dissolve. Add the grape juice concentrate and stir in the lavender flowers, the acid blend and yeast nutrient.

2. Pour into a fermentation bucket and add the rest of the water. Cover and leave overnight. Take a hydrometer reading now if you wish.

3. Sprinkle on the yeast and cover the bucket loosely, allowing fermentation to start. Stand for 5 days and then rack off into a demijohn, and fit an airlock.

4. Allow to ferment out over several weeks or a few months. If you wish the finished wine to be sweet rack off at 3 weeks and add 1 crushed Campden tablet. Leave overnight and then sweeten with sugar to taste.

5. If you wish the wine to be drier, rack off when fermentation has completely finished. Fill the bottle to the neck with water and leave to settle before bottling. Store for at least 6 months to mature.

LEMONS

Limoncello

Ingredients:
5 unwaxed lemons
1 750 ml bottle vodka
750 g sugar
700 ml boiling water

Method:

1. Pare the zest from the lemons and put in a large screwtop jar with the vodka. Secure the lid shake, then store in a cool, dark place for a week, shaking daily.

2. In a pan, boil 700 ml of water and add the sugar, stirring till it dissolves. Cool to warm and then pour this sugar mix into the large screwtop jar containing the vodka/lemon mix. Shake daily for another 7 days.

3. Strain through muslin into bottles adding a few strips of lemon zest.

NASTURTIUMS

Nasturtium Vodka

This is a flower infusion of a base spirit. The recipe is here for this, but you can substitute gin, eau de vie or rum for your base and you can use a variety of edible flowers, spices or fruits to infuse. The nasturtiums will colour the vodka and add a peppery bite to it.

Pick enough nasturtiums to fill a screwtop jar, shake any beasties off the flowers and place in the jar.

Pour over with the vodka, screw on the lid, shake and then leave in a cool, dark place for two days. Strain off the liquor into a clean bottle. Can be drunk immediately.

NETTLES

Nettle Wine

Ingredients:
1.5 litres of nettle tops (loosely pressed down to measure)
1 kg sugar
1 litre white grape juice
juice of 2 lemons
a small piece of root ginger, bashed (optional)
1 cup of strong black tea
1 sachet of wine yeast
yeast nutrient (optional)
Water to 1 gallon

Method:

1. Wash the nettles well in cold water, place in fermentation bucket, and add 1 pint cold water. You can add 1 crushed Campden tablet at this point if you are cautious.

2. Leave overnight and then stir well. Add the grape juice, lemon juice and black tea now, stir well.

3. In a pan boil a pint of water and dissolve the sugar in it. Add the peeled bashed ginger is you are using it. Allow to cool a little and transfer into the fermentation bucket.

4. When the temperature of the liquid has reached room temperature, then sprinkle on the yeast. Cover lightly with the lid and leave to sit for a week, stirring daily to push the nettles into the liquid.

5. Strain through a straining bag or muslin into a

demijohn, adding more water if required to take the liquid level up to the shoulders (you can take an initial S.G. reading at this point, which should be around 1.090-1.100). Fit an airlock and leave to ferment out. Take further S.G. readings till you get to a reading of 990.

6. Rack into a clean demijohn, fill to neck with water if required, fit a new airlock and leave to mature for 4 months, then bottle.

OAK LEAVES

Oak Leaf Wine

This makes 6 bottles.

Ingredients:
About 50 g of washed fresh oak leaves.
3 cm long piece of fresh ginger root
2 kg of sugar
500 g chopped raisins
1 sachet white wine yeast

Method:

1. Place your clean fresh oak leaves in a plastic fermentation bucket. Cover with enough boiling water to submerge the leaves. Cover loosely and allow to stand for five days.

2. Strain through a straining bag and conserve the liquor. Pour this into a large pan and heat, adding the ginger and the sugar and stirring to dissolve. Bring to a simmer for 30 minutes.

3. Remove from the heat and add the chopped raisins. When cool, sprinkle on the yeast.

4. Transfer to a demijohn, and fit an airlock.

5. Ferment for 2 weeks. Strain through a straining bag, and into a clean demijohn, filling to the neck with water. Fit a clean airlock and leave to stand for a week checking if there is further fermentation.

When completely fermented out, taste and if it requires further sweetening you can add 1 crushed Campden tablet, leave overnight, then bottle the next day after adding sugar to taste.

PARSNIPS

Parsnip Wine

This is one of the oldest recipes, having been made since earliest times. The parsnip, like the carrot is a root full of sugar, which sweetens considerably after a frost. This wine has a good complex taste and is a little like sherry or Madeira. Use the parsnips after they've been frosted – it makes the plant convert the starch into sugar.

Ingredients:
2 kg parsnips
1.5 kg sugar
4.5 litres of water
1 sachet white wine yeast
Juice and zest of 2 oranges (optional)
You could add a clove or two or
pinch of cinnamon

Method:

1. Peel and chop the parsnips, and boil in the water in a large pan for 15 minutes. Add any spices now (and the orange is using) and cook gently for a further 10 minutes.

2. Strain into a clean bucket, add the sugar, stir to dissolve and leave to cool.

3. Sprinkle on the yeast and cover and let it sit for 2 days. Then transfer into a demijohn and fit an airlock.

4. When fermentation has totally finished, rack off into a clean demijohn, top up to the neck with water and fit a clean airlock.

5. Leave to mature for at least 6 months, then rack into bottles. Store for a further year before drinking, but improves up to 2 years.

PEAS

Peapod Wine

This is one of my favourite wine made from vegetables and is very eco-friendly. You can grow peas, shell them and eat the peas then make wine from the pods, and then after wine-making you can feed the spent pods to any farm animals you have. They in turn convert it to manure for the garden. This makes a dry white wine.

Ingredients:
2.5 kg peapods (collect them and freeze them till you have enough)
200 g sultanas
800 g sugar
Pectolase
1 cup black tea
1 sachet white wine yeast
1 tsp acid blend
2 litres water

Method:

1. Chop the sultanas and add to the peapods in a large plastic bucket with 2 litres of boiling water.

2. Leave to soak for half an hour, then strain off into a clean bucket. Add the sugar, stir well and leave to cool.

3. Add all the other ingredients and stir, then finally sprinkle on the yeast. Pour into a demijohn, top up to the shoulder with water if necessary and fit an airlock. Leave to ferment in a warm place.

4. After a week or so, the initial fermentation will have slowed down. You can gently agitate the liquid in the demijohn, and then after a second week, rack off into a clean demijohn, topping this up with water to the neck and fitting a new airlock.

5. After a month or so fermentation should have completely ceased, and you can siphon off into bottles, filling to leave a 5cm headspace. Mature in a cool place for 6 months minimum.

PEARS

Pear Wine

Ingredients:
5 lb of pears
1 kg sugar
250 ml white grape concentrate (or grape juice)
1 tsp citric acid (or lemon juice)
2 tsp pectolase
1 sachet white wine yeast
1 tsp yeast nutrient
1/2 tsp potassium sorbate
1 crushed Campden tablet.

Method:

1. Wash and chop the pears and transfer them and any juice to a pan and boil for 10 minutes in a gallon of water.

2. Pour through a straining bag into a fermentation bucket on top of the sugar, and allow to cool.

3. Add the grape concentrate, the pectolase, the yeast nutrient and the citric acid and stir to mix.

4. Sprinkle on the yeast, and loosely cover with the lid, and leave to sit in a warm place for 24 hours.

5. Pour into a demijohn and fit an airlock, and allow to ferment out. When fermentation has ceased rack off into a clean demijohn, leaving the lees behind. Add the potassium sorbate and 1 crushed Campden

tablet. Allow the fizzing that may occur to escape, top up to the neck with water then fit a new airlock.

6. Move the demijohn to a cold place to clear and settle and when clear bottle.

7. If you need to further clear the wine you can add finings before bottling. If you need to sweeten it use a non-fermentable sweetener like Splenda.

Perry

True perry is made from perry pears, as opposed to pear cider, which can be (if you're lucky, apple cider flavoured with pear juice, and if you are unlucky a chemical brew consisting mainly of water, sugar syrup and lots of colourings and flavourings). Perry pears are astringent and hard and like apple cider takes many months to mature properly to make a decent drink. Furthermore, they have to be milled and pressed at just the right time; not quite ripe means there is less sugar available to ferment, too ripe and they clog up the mill and press. The best British perry maker in my opinion is Tom Oliver who lives and works in the great perry growing area of Ocle Pychard (yes really) in Herefordshire, England. I've met Tom and visited his farm and you really couldn't meet a nicer man. He introduced me to perry (I was a disbeliever), and changed my opinion immediately. A good perry should rival a good white wine, and his was sublime.

You will need real fruit for this as pear juice just doesn't really work. 10 kg of pears will make about a gallon of perry. Dessert pears will go very soft when ripe, so if you can use perry pears or wild ripe pears. I don't make perry with my rack and cloth press as I don't have perry pears and cannot extract the dessert juice easily from the cloths, but have tried wild ripe pears which are easier.

Ingredients:
10 kg perry or mixed pears
1 cup black tea
1 sachet champagne yeast
1 crushed Campden tablet
Pectolase

Method:

1. Pulp and press the pears as you would for cider, and when you have the juice place in a fermentation bucket.

2. Add 1 crushed Campden tablet for every gallon of juice and leave overnight to kill any bacteria and wild yeast. You can also add pectic enzyme (pectolase) at this stage (1 tsp per gallon), and a cup of black tea to add some tannin notes.

3. Transfer to a demijohn, sprinkle on a sachet of champagne yeast and fit an airlock.

4. After about four days syphon off the perry into a clean demijohn and top up with water to the neck. Fit a clean airlock and leave to fully ferment out which should take a couple of weeks.

5. The pectolase should help clear the perry, but placing the demijohn into a cold area the night before you rack off should also help the yeast drop out of solution. You can then bottle.

6. Now, if you are making this in say September time, you will be able to drink your bottled perry by Christmas, but if you place some of the bottles in a cold garage or shed over winter, they will be even better to drink by Easter.

PINEAPPLE

Pineapple Wine

This makes a slightly hazy wine. You can also use this recipe and substitute mango, guava or peaches. Will be ready to drink in 5 days but better after maturation for a month. The egg-white clarifies this wine, which although sweet has a little bit of bitterness. You can increase the quantities to make a full demijohn, but I'd try this small quantity first.

Ingredients:
500 g pineapple (can be fresh or tinned in juice)
500 g sugar
1 sachet white wine yeast
2 tbsp sprouted or whole wheat grains. (sprouted will ferment faster)
500 ml water
1 cardamom seed
1 small piece of cinnamon stick
2 cloves
1 tsp egg white

Method:

1. Firstly, if you have a fresh, ripe pineapple, peel and core it and cut into small chunks. If you are using tinned pineapple in juice, drain off the juice and set aside.

2. Put the pineapple pieces in a mixing bowl and grind into a paste with a wooden spoon or meat tenderiser. Then empty this paste and any juice into a clean large screwtop jar.

3. Add the sugar, the spices, the wheat and the egg white. Then pour on the 500 ml of warm water. Ensure you leave about a third of the jar empty for the fermentation to fill. Finally, sprinkle on the yeast and put the lid on, tightly. Shake for a minute or so to mix the ingredients and dissolve the sugar, then loosen the lid slightly and place the jar in a cool, dark place to ferment.

4. Turn the jar daily, ensuring you first tighten the lid, then loosen slightly again before replacing to ferment. Do this for 5 days.

5. Strain through a straining bag into a clean wide mouthed bottle and leave to clear in a cold place for a day or so. Then you can siphon the wine off any lees into a clean bottle. Ready to drink now, but better in 3 weeks.

PLUMS

Plum Wine

Ingredients:
2 kg plums or mirabelles
1.5 kg sugar
1 sachet wine yeast
1 gallon water

Method:

1. Cut up the plums and remove the stones. Place in the fermentation bucket and pour on the gallon of boiling water and stir.

2. Leave covered for four days, stirring daily.

3. Strain the fruit out and discard, leaving the liquor in the bucket. Stir in the sugar to dissolve and then sprinkle on the yeast.

4. Leave overnight and then transfer to a demijohn fitted with an airlock to ferment. Place demijohn is a warmish place.

5. When all fermentation has ceased, siphon off into a clean demijohn, top up to the neck with water, fit a new airlock and leave to mature for 6 months, before bottling.

POTATOES

Potato Wine

Ingredients:
1 kg potatoes
450 g pearl barley (optional)
1 kg sugar
225 g chopped sultanas or a 250 g
can of white grape juice concentrate
1 tsp acid blend
Juice of 1 lemon or orange
1 gallon water
1 sachet white wine yeast

Method:

1. Wash the potatoes thoroughly and chop them but do not peel. Place them in a large saucepan with the pearl barley and simmer until the potatoes are tender.

2. Strain through a straining bag or muslin into a fermentation bucket on top of the sugar and stir until the sugar has completely dissolved. When cool add the chopped sultanas, lemon/orange juice, acid blend and wine yeast and cover the bucket and leave for five days.

3. Strain into a demijohn, fit an airlock and leave to ferment.

4. When fermentation has ceased, rack the wine into a clean jar and place in a cool, dark place and leave for a further few months. Rack again if necessary and leave until the wine is clear and stable and then bottle. This wine goes on improving for well over 12 months.

RASPBERRIES

Raspberry Gin

Ingredients:
350 g raspberries (or gooseberries,
plums, strawberries or currants)
150 g sugar
700 ml gin

Method:

1. Tip the clean raspberries and the sugar into a 1.5L sterilised jar with lid. Add the gin, seal the jar and *gently* swirl the contents around to dissolve the sugar. Don't shake as you will break up the delicate fruit. Store in a cool, dark place, gently turning the jar daily for 7 days. After 14 more days, strain the gin through a muslin and bottle. Fruit can be used in a boozy trifle.

Raspberry Wine

Ingredients:
2 kg raspberries
500 g sugar
1 vitamin C tablet (or some rosehip juice)
1 sachet wine yeast

Method:

1. Place the fruit in a muslin or straining bag and secure in a fermentation bucket.

2. Add a kettleful of boiling water and mash the fruit inside the bag with a potato masher.

3. Lift the bag and the contents out and discard. Strain the juice through a second clean straining bag into a clean bucket. Allow to cool.

4. Heat 1 litre of water in a pan with the sugar, stirring to dissolve. Allow this to cool and add to the juice bucket. Stir to mix and then transfer into a demijohn, topping up with tepid water if necessary.

5. Sprinkle on the yeast, fit an airlock and allow to ferment. When you think the fermentation is finished carefully move the demijohn to a cold place overnight to allow the yeast to settle out of solution. You can add a crushed Campden tablet at this stage, then carefully rack off into bottles. This makes a nice light wine.

ROSES

Roses are part of the same family as apples, and any product you make with them may have a slightly delicate scent of apples.

Rose Petal Vodka

This is a delicately flavoured very easy and quick summer vodka and can use any rose petals.

Ingredients:
Couple of handfuls of fresh clean rose petals
Half bottle of vodka

Method:

1. Take a good handful or two of scented fresh rose petals and push them into a sterilised wide-mouthed large jar (not a bottle), and pour your chosen vodka on top, tighten lid and leave in a dark cool place to infuse for 2 or 3 days.

2. Strain through a muslin or paper coffee filter into a clean bottle. Taste and if not sweet enough add a little sugar prior to serving.

Rosehip Wine

This wine is a warmly coloured, and has a strong tannin taste, which can be artificially sweetened when complete. You can use fresh, dried or frozen rosehips. Remove the woody end of the rosehip and de-seed to reduce the tannins. Freezing the prepared hips will soften the flesh prior to fermentation.

Ingredients:
1 kg fresh or 250g dried rosehips
4.5 litres water
1 kg sugar
Juice of 1 lemon or orange (optional)
1 sachet red wine yeast
1 tsp pectolase
1 tsp yeast nutrient
1 Campden tablet.

Method:

1. Wash and bash your rosehips and place in a secured straining bag at the bottom of the fermentation bucket.

2. In a large pan, heat half the water (2 litres) and all the sugar, stir to dissolve, then boil. Pour the boiling water into the fermentation bucket, covering the bag containing the fruit.

3. Add the remaining water (2.5 litres) and the lemon or orange juice, and then add the crushed Campden tablet, and leave to sit for 2 days.

4. Add the pectolase, yeast nutrient and stir. Use the hydrometer to take an initial S.G. reading. Then

sprinkle the yeast on the surface of the liquid. Fit the lid and add an airlock and leave for 7 days to ferment.

5. Remove the straining bag, allowing the liquid to drain back into the bucket. Re-fit the lid with airlock and leave to sit for one or two days.

6. Rack into a demijohn fitted with an airlock, filling to 5cm from the top. Leave for 3 months to finish fermentation and clear.

7. If there is any new sediment, rack off again into a clean demijohn, filling to the top, and take a final S.G. reading. Leave for a further 3 months and then bottle. Best after at least three further months but will continue to mature.

RHUBARB

Rhubarb Gin

Ingredients:
1 kg pink rhubarb stems
400 g caster or granulated sugar
800 ml gin

Method:

1. Wash and trim the rhubarb to use only the pink bits of stem and cut into 3 cm sticks. Place in a large screw top jar with the sugar, tighten the lid and give it a shake. Then leave overnight to allow the sugar to draw the liquid and colour from the rhubarb.

2. Next day, add the gin, reseal the jar, shake well and put in a cool dark place for four weeks, shaking occasionally. Strain the liquor through a muslin and bottle it. The pink colour will fade.

Rhubarb Wine

This makes a nice rosé wine with a bit of a kick.

Ingredients:
1.5 kg rhubarb
1.3 kg sugar
50 ml white grape juice concentrate
1 sachet white wine yeast
1 tsp yeast nutrient

Method:

1. Choosing the pinkest stems of the rhubarb, wash and remove leaves and the white part at the start of the stem. Chop the stems into 2 cm thick slices. Discard the green slices, and put the pink slices into your fermentation bucket.

2. Add the sugar, mix well and cover and leave to sit for 3 days. Then pound the fruit with a potato masher and stir in 3 litres of water.

3. Strain the liquor into a clean bucket and add the grape juice concentrate.

4. Pour into a demijohn, top up to the shoulders with water, add the yeast and fit an airlock.

5. Leave to ferment for 10 days, then when the fermentation slows down, top up the demijohn to the neck with water, and refit the airlock.

6. When fermentation has completely finished rack off into a clean demijohn, fill again to neck, fit a clean airlock and leave in a cold place to clear for 2 days and then if no further fermentation has occurred syphon off and bottle.

RICE

Rice Wine or Saki

Using Basmati rice will create a stronger wine than other white rices. The lemon juice is added to increase the acidity in the wine and also to give a little flavour complexity. If you can find Kobi rice, you can make a more authentic Saki wine, but this recipe is a good starter.

Ingredients:
1 kg basmati rice
250 g white grape concentrate
1 kg sugar
2 tsps acid blend
Squirt of lemon juice
1 gallon water
Yeast Nutrient
1 sachet white wine yeast
1 Campden tablet

Method:

1. Rinse and coarsely crush rice. Place rice in a straining bag and secure the top and place in fermentation bucket.

2. Pour hot water over rice and stir in all ingredients except yeast and nutrient. Cover with lid and leave to sit for 48 hours.

3. Add the yeast nutrient and sprinkle on the yeast. You can take an Initial S.G. reading at this time.

4. Stir daily and allow to ferment for a couple of days and check the S.G., which should be around 1.050.

If not add in some sugar, stir to dissolve and take another S.G. reading. Cover with the lid again.

5. Take a further S.G. reading at 7 days, when the S.G. should have dropped to about 1.030. If it is over this, leave for a further day and test again. Remove the bag and allow the juice to drain back into the bucket. Pour this into a demijohn and fit an airlock.

6. Test the S.G. again about a week later. When it has dropped to 1.010 syphon the wine into a clean demijohn, leaving the lees behind. Take the opportunity to taste for sweetness. You can add a crushed Campden tablet to stabilize the wine and then fit a clean airlock. Leave overnight, and the next day you can add a little sugar if you feel the wine needs sweetened, refitting the airlock and swirling a little to mix. Remember allowing oxygen in may destabilise the wine.

7. If the airlock has ceased to show any bubbles at all, you can move the demijohn into a cold room and leave overnight to help clear it. Then siphon into bottles. Can be drunk after a week or so, but better if matured for a month.

TEA

Yes, you really can ferment different types of tea to make alcohol. Tea leaves are grown on the bushes of the *Camellia sinensis* plant and the top two leaves and bud only are removed and processed.

Fruity tea-bag wine

Ingredients:
20 supermarket Fruit Tea Bags
1.5 kg Sugar (for a dryer, less sweet wine use 1.2Kg)
1 sachet wine yeast

Method:

1. Add 4 litres of water to a large saucepan and bring it to the boil. Turn off the heat, add the sugar and stir until dissolved.

2. Add the teabags and gently stir (do not break the teabags). Put the pot lid on and leave to steep until cool.

3. When cool, gently remove the teabags with a large spoon. Pour into a demijohn, top up with water if required to the shoulders, sprinkle on the yeast and fit an airlock.

4. Move to a warm area where the demijohn can sit undisturbed to ferment for around 10 days.

5. If fermentation has ceased (no bubble activity in the airlock), rack off into clean demijohn, top

up with water to neck and fit a clean airlock. This should allow the wine to fully clear in a week or so, and then you are ready to siphon into bottles. If you wish you can add 1 crushed Campden tablet to the demijohn the night before you bottle to stabilise it. If you wish the wine to be a little sweeter then you can add 1 crushed Campden tablet just before the first racking off.

Tea Wine

This does actually make a nice wine (that doesn't taste of tea).

Ingredients:
4 litres of strong cold black tea
2 kg sugar
500 g raisins
3 lemons
1 sachet of wine yeast

Method:

1. Cut up the raisins and slice up the lemons and put them in a fermentation bucket. Add the sugar and pour on the strong cold tea. Stir until the sugar dissolves, cover and leave in a warm place for a month to ferment.

2. You should see a scum-like layer on the top of this wine. It is difficult to remove but you can siphon off the liquid below this, being careful not to disturb the sediment. Add one crushed Campden tablet before bottling and bottle immediately. It can be drunk now, but will mature and taste better after 2 months.

WALNUTS

Eau de Noix

The traditional day to make this is on the 24th June; St John's day. The walnuts must be picked before July 14th and be green. It will be ready to drink 100 days later. This makes a litre, and is reminiscent of port.

Ingredients:
500 g brandy
500 g strong red wine (corbieres or some other wine from south of France)
500 g sugar
2 cloves
1 vanilla pod
Small piece cinnamon stick.

Method:

1. Place the shelled and quartered green walnuts in a large wide-mouthed screw top jar with the spices.

2. Pour the brandy and wine over the walnuts and secure, turning to mix daily for a week.

3. Store in a dark cool place turning weekly for 100 days.

4. Filter and add the sugar before bottling. For a long shelf life you can add 1 crushed Campden tablet at this time, otherwise keep refrigerated and consume within a month.

INFUSIONS

You can make an infusion with almost any base spirit – vodka, gin, rum, whisky, eau de vie, ouzo, and rum, and using a variety of ingredients. Some will taste nice than others, and the most bland base spirit is vodka, so it is a favourite to use.

I've listed some thoughts below- they're not really recipes, but suggestions.

Pink Gin

Use cheap dry gin as a base. Pour into a screwtop jar with a handful of raspberries, some pink rose petals, a drop of angostura bitters if available, or a small amount of pomegranate juice. Some orange peel or lemon zest is acceptable. Leave for a week, then strain out.

Gin Infusions

Rhubarb - chop pink rhubarb into 1 inch long stalks, infuse for 4 weeks.

Raspberry - 350g raspberries infused in gin for 2-3 weeks.

Plums - Cut in half 300g of plums or damsons infuse in a bottle of gin for a month

Elderflower & apple gin - elderflower florets, lemon peel & slices of apple in gin to infuse for 2 days.

Vodka Infusions

Ginger - lemon peel, peeled bashed ginger & mandarin peel in vodka- infuse for 1 day.

Cherries - put whole cherries in vodka in a jar. Infuse for 3 weeks.

Chilli vodka – put ONE chilli (halved and de-seeded) into a jar with vodka. Remove 3 days later.

Brandy Infusions

Brandy usually used but port or vermouth can be used.

Infuse with any or all of the following -

Juniper berries, vanilla, orange peel, tea, cloves, star anise, coffee beans, saffron, mace, nutmeg, cinnamon, lemon balm, thyme, bay, rosemary, honey, Artemisia (mugwort), oregano, fennel, camomile & elderflower.

Schnapps is a fermented fruit liquor usually made in Germany; in France it's known as Eau de Vie. In Europe it's frequently drunk straight as an aperitif (being somewhat dry), whilst in the UK and USA it's sweeter and used as a base for a cocktail. As this is a distilled alcoholic drink, I won't be explaining how to make this, but can make suggestions for you to flavour your bought Schnapps or Eau de Vie.

Raspberry is probably the favourite flavour, although peach, plum, cherry, orange and ginger and nasturtium are interesting to try. You simply get a screwtop bottle or jar, add your chosen flavouring and place inside and pour on the Eau de vie, Schnapps or vodka. Shake gently, leave in a cool dark place for a week, then strain the liqueur out and rebottle.

PASTEURISATION & STORAGE

Naturally, once we have made our beer, wine, cider or liqueurs we want to drink them, but if we have got a favourite recipe and make an amount that cannot be drunk immediately or we want to lay some down for future enjoyment, then we need to store it.

There are a couple of methods of ensuring that what we make remains stable, including pasteurisation, fermenting to dryness or killing the yeast, so we will start with pasteurisation.

PASTEURISATION

Pasteurisation uses prolonged heating of a fluid to destroy the yeast cells within it, rendering it unable to multiply. This stops the liquid fermenting further and stops any dangerous build up of Carbon dioxide gas within a container. Normally cider is bottled or stored in bag in box bags, which have been produced to withstand the pasteurisation process. Either can be pasteurised in a bath pasteuriser for 20 minutes at 70 °C, which will be sufficient to kill any yeast within. This process also kills most bacteria. A bottle uncapped and with a thermometer placed in the water will allow you to monitor the temperature if you do not use a self-regulating pasteuriser. Once you remove the bottles or bags from the pasteuriser you can lay them on a clean towel to cool naturally, the liquid

183

moving into the neck and cap of the bottle or into the tap on a BIB bag will pasteurise inside this.

You cannot pasteurise normal plastic bottles or recycled plastic milk containers – they become unstable in this heat. If you pasteurise at a lower temperature or for under 20 minutes you risk not destroying the yeast cells and any bacteria therein. Any higher than 70 °C and you will cook the cider, resulting in a 'toffee apple' taste.

You can find pasteurisers from specialist suppliers such as Vigo in Devon (see the appendix for details).

If you decide to pasteurise apple juice, remember that the process will kill off *some* bacteria but not all (hence the guidance on picking clean, undamaged apples). There is a nasty pathogen called Patulin which is not destroyed by pasteurisation (but is destroyed by fermentation). Patulin can give you an upset stomach, vomiting or diarrhoea.

FERMENTING TO DRYNESS

If we take cider as an example here, fermenting to dryness means fermenting all the available sugar in the juice and turning it into alcohol. This resulting 'dry' cider is stable enough to be bottled without pasteurisation, or can be bottle conditioned with a very small charge of sugar before capping to start a small amount of fermentation in the bottle resulting in a naturally carbonated cider. The teensy issue is that dry cider; whether it is sparkling or not is not everyone's cup of tea. My son, for example likes a sparkling medium cider. If this is your bag then you can either add a non fermenting sugar (Splenda is excellent – one tiny tab is enough for a 500 ml bottle) or you can stop your cider fermenting *before* you get to dryness. This isn't easy and not for the beginner. Even further down the complicated and experienced journey of cider making is keeving; which produces a naturally sweet, sparkling cider. I only briefly explain it in this book as it is beyond the scope of most beginners.

Fermenting a beer to dryness means you can also bottle it safely for storage purposes either in glass bottles or in a plastic beer barrel. Store in a cool dry place away from strong smells or chemicals.

Wine can be bottled when fermentation is complete and stored in wine bottles in a cool, dark place.

Liqueurs are normally stable once the fruit that you have added to colour or flavour the spirits is removed, the alcohol in the spirit being so high as to prevent any further fermentation.

KILLING THE YEAST

Again, taking cider as an example, when we 'kill off' the yeast *before* fermentation is finished we can stop the cider. Sometimes in cold climates the fermentation process naturally slows right down due to cold temperatures in the winter or when the sugar in the liquid has all been turned to alcohol, but the yeast is still alive; it's just dormant. If we wanted to stop the process *with certainty*, we can either pasteurise or chemically kill the yeast by using Campden tablets.

This is why a notebook marking down the fall in S.G. rates is important. A little taste as you near the magic reading of 1.015 or 1.010 will determine where you stop it. Your own personal preference for sweetness is very individual. One man's sweet is another's medium – you need to taste it. Stopping the cider early with this method and bottling it may result in a very small residual amount of carbonation, but most likely it will turn out still; unless you prime individual bottles with a very small amount of sugar, fill with the cider and cap immediately; allowing a small amount of carbon dioxide to be produced by the live yeast. Bottle conditioning needs to be done with care, see the instructions for cider and beer in those sections.

QUICK CHEAP BOOZE

Well, welcome to the fast and furious section of the book! The recipes here are for quick and cheap alcohol.

ALCOHOLIC GINGER BEER – Ready in 48 hours!

Firstly let's clear up a bit of confusion – ginger ale, ginger beer and alcoholic ginger beer are three different things. Ginger ale is basically a soft drink made of water with ginger flavouring. Ginger beer is a similar thing – non alcoholic and used as a mixer for strong spirits. Alcoholic ginger beer is a whole different animal. You either like it or hate it; it's a kind of Marmite thing; but one thing is for sure, if you've tried real alcoholic ginger beer, you will never forget it. This is not a subtle drink – whether you sweeten it (advised) or like it dry, that first sip will assault your tongue and inform you just how spicy raw ginger really is!

Ingredients:
75 g finely chopped or grated fresh
raw ginger root
2 unwaxed lemons, sliced thin
500 g golden caster sugar
1 tsp cream of tartar*
Sachet of dried wine yeast (white
wine or Chardonnay)
4.5 litres water
*You can substitute Baking soda (1.5
tsps for 1 tsp cream of tartar)

Method:

1. Put the ginger, lemons, sugar and cream of tartar and
   1.5 litres of water into a large pan and heat gently heat,
   stirring to dissolve the sugar.

2. Bring to boil, turn off heat and then add 3 litres of
   cold water. Stir to mix. Pour into fermenting bucket.

3. Stir in the sachet of yeast, cover bucket loosely with
   lid and leave overnight in a cool place.

4. Strain through muslin or clean paper coffee filter
   and then pour into sterilised plastic screw top
   bottles (recycled water or pop bottles), leaving a
   space of about 5cm at top. Screw lids on tightly and
   store in a cool place, unscrewing tops slightly to
   release gas and pressure every few hours.

5. This is ready to drink in 48 hours. Keep chilled and
   consume within 3 days. Don't forget to release the
   gas pressure if the bottles feel very rigid – this is
   very important! If it's not sweet enough for you
   then you can add some Splenda or other artificial

sweetener – if you add sugar at this stage it will re-ferment. If you don't think it is alcoholic enough then you can pour it back into the bucket, add some more sugar and re-ferment it for a couple more days.

6. If you wish to keep it much longer then you can kill the yeast by adding Campden tablets (check the dosage) or put into glass bottles with crown caps and pasteurise for 20 minutes at 60 °C. It will no longer be sparkling, but still.

TURBO CIDER – Ready in 1 week!

Yes, I've made it, and yes I've drunk it. It has to be one of the simplest and cheapest ways to make anything alcoholic. It's an acquired taste, but popular amongst students and those who cannot afford proper cider making equipment, and it will be ready to drink in 8 days. Now, the more astute reader will no doubt be saying to themselves – *hang on a minute, this is just fermented fruit juice, I could do this with cranberry juice, pear juice, pineapple juice…'* And of course, astute reader you can. My job in this book is not just to educate and give you confidence, but to inspire you to try things out and experiment. Now, your experimental booze making may not always turn out to be great drinks, but you'll have tried it and maybe learnt something. The only caveat I would say is a fruit juice wine needs to mature for at least 6 months, whereas this turbo cider recipe can be drunk very quickly.

Ingredients:
5 litres of supermarket apple juice
(without preservatives is best).
1 sachet cider yeast

Method:

1. Pour 4.5 litres of apple juice into a demijohn, filling to the shoulders.

2. Sprinkle the yeast on, fit an airlock, and place in a warm place for 7 days. Fermentation will be quite lively, and then calming down.

3. When the bubbling has ceased, rack off into a clean demijohn.

4. NOW - you can drink this now – it's young and a bit rough, but for me, when you rack off, top up to the neck with water (you MUST keep the air out of it), fit a clean airlock and leave it in a cool place to mature. Even a fortnight will improve the taste, and allow the cider to clear.

5. If you then wish to make it slightly sweeter, add 1 crushed Campden tablet to the demijohn, leave overnight and back-sweeten with apple juice, or an artificial sweetener.

6. If you wish to bottle condition, when you are sure the cider has completely fermented to dry prime the bottles with 1/4 tsp of sugar before you fill with cider leaving a 5cm headspace. Glass bottles with crown caps. Store in a cold place for at least 10 days.

DANDELION BEER – Ready in 1 week!

I actually really like this recipe, and so does everyone who's tried it. The preparation makes it look a little foul, but as a cold summer drink it's refreshing and moreish.

Ingredients:
225 g whole, young dandelion plants
450 g Demerara sugar
5 g root ginger
1 large lemon
25 g cream of tartar
1 sachet white wine yeast
1 gallon water

Method:

1. First gather your dandelions. Dig up the whole plant, keeping the roots intact. Big fat roots are best, so choose the biggest plants. Wash thoroughly to eliminate all soil, and remove any threadlike roots. Peel and chop the ginger, and peel the lemon.

2. Put the entire plants into a large saucepan with the water, ginger and the peel of the lemon. Bring to the boil and simmer for 10 minutes.

3. Strain through a bag or muslin into the fermentation bucket. Add the sugar and the cream of tartar and mix thoroughly. Allow to cool to room temperature.

4. Add the lemon juice and stir in, then sprinkle on the yeast. Cover loosely with the lid and leave to stand for three days, stirring gently once a day.

5. Fermentation is fast. Strain again into a clean

container, leave to settle for a few hours and then syphon and bottle into plastic screw top bottles.

6. Store in a cold place, and religiously check the rigidity of the bottles a least once daily, and release the pressure slowly if required. Ready to drink in one week.

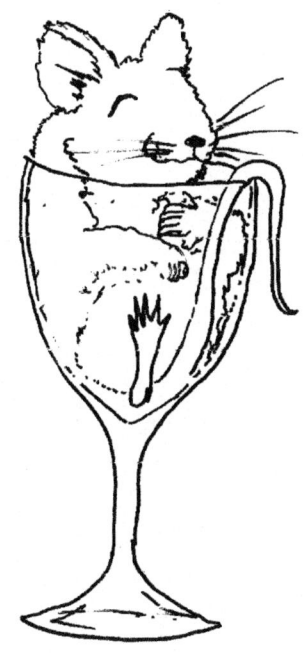

TINNED FRUIT WINE – Ready in 6 months.

Yes, the humble supermarket tin of fruit in syrup or juice can also be transformed into cheap wine.

The main difference between using tinned fruit as opposed to fresh, is that tinned fruit in a sugar syrup will not need as much sugar to ferment. If the tinned fruit is tinned in water or fruit juice, then use the same amount of sugar as you would for fresh fruit. Tinned apricots, pineapple, pears or peaches are a good choice, and can be mixed. Avoid fruit salad, as this nearly always contains preservatives and won't ferment. If in doubt, always check the ingredients on the label. Anything that contains Ascorbic acid, sulphur dioxide (sulphites), benzoate or potassium sorbate won't ferment.

Ingredients:
2 or 3 x 400 g tin of fruit
1 kg sugar
1 ½ tsp citric acid (or a good squirt
of lemon juice)
1 cup of strong black tea
Pectolase
general-purpose wine yeast
Yeast nutrient (optional)
Water to 1 gallon
Optional- A small can of Red/White
Grape Concentrate. It will add body
to the wine, or a small handful of
raisins.

Method:

1. Open the tin and pour the entire contents into the fermentation bucket. Mash the fruit with a potato masher.

2. Heat 2 litres of water in a pan and dissolve the sugar in this, then add it to the bucket.

3. Allow to cool to room temperature, add the citric acid, black tea, pectolase, grape concentrate (if used), yeast nutrient, and finally sprinkle on the yeast.

4. Stir daily for four days and then strain wine into a demijohn through a nylon sieve and fit an air-lock. Ferment out, rack into a clean demijohn, add one crushed Campden tablet to stabilise and sweeten with sugar dissolved in water if an off-dry wine is preferred. Rack again after a month and leave to mature for three months, rack again if sediment builds up. Bottle and leave for at least six months before drinking.

APPENDIX

Ice Cubes

A tiny word about ice-cubes. If you'd like to make some attractive ice cubes with a difference, then boil and cool the water before filling your ice cube tray. This will result in clear cubes with no air bubbles. If you'd like to further enhance your drinks, then adding a few edible flowers to the water prior to freezing will make pretty ice cubes. Borage is perfect for ice cubes for Pimms, elderflower florets for ice cubes for gin and cocktails. If you make ice cubes from frozen elderflower cordial these are perfect in a gin and tonic.

Frugal rack and cloth cider press plans.

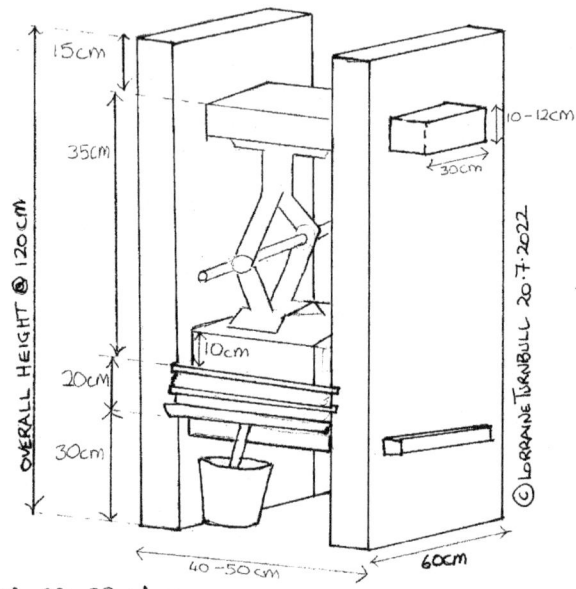

FORMER 30 x 30 x 4cm
PRESS RACKS 30 x 30 cm
JUICE TRAY 40 cm²

If you are short of cash and want to make your own wooden cider press, check out this idea. The manual car jack is much quicker than a manual screw and less fraught with danger than an overhead hydraulic jack, which could leak hydraulic fluid into the cider. I have indicated sizes on the drawing, but most self-sufficient and frugal folk will want to use what they have when possible. Use solid timber, not MDF or plywood.

You will need:
A car jack
A platform of some sort with a solid
back (a table against a wall is good)

Some square pressing cloths – you can use polyester net curtains
Some square or large rectangular polypropylene boards
A stainless steel or plastic former
A stainless steel or plastic juice collecting tray with a hole front and centre
Length of food-grade clear plastic hose
Timber as per plan measurements

How to make it:

So you'll need to make a frame from decent timber. The drawing shows you basically have two tall upright rectangles of wood (these could be ready-bought solid wooden shelves). You will need to cut holes in the top section of these uprights in order to house a solid piece of timber inside. This needs to be able to take the top of the jack underneath and thick enough to take the amount of pressure required to press the racks below.

The uprights could be between 100-140 cm tall, if you are making this with a bottom recess to site a collection bucket; or just shorter if you are going to use this on a solid surface. If the latter then you'll need to move the collection hole to the front centre of the tray, not underneath front and centre.

At the bottom of the press, you can either fit a secure and strong base of solid wood, or rest the uprights onto a flat solid surface – perhaps a kitchen worktop or on top of breeze block bricks outdoors. Remember you will need to be able to fit a short piece of hose from the hole in the juice tray into a receptacle to catch the juice.

Then from the bottom upwards you place your plastic or stainless steel juice tray (with the hose fitted tightly to allow the juice to flow into a bucket or jug).

Inside this place your former. This should be a rectangular or square hollow plastic or stainless steel mould (about 30 cm square and about 3-4 cm deep).

Inside the former lay your first square of pressing cloth in a diamond shape over the square former. This needs to be sufficiently big to be able to lay inside the former full of apple pulp and then fold over the excess (like an envelope) fully containing the pulp in the fabric. This is called a 'cheese'.

On top of this place a pressing rack. This can be as simple as a plastic food chopping board.

Then put the former back on and make another cheese and place on top another pressing rack.

For a simple press like this, two or three cheeses are plenty. Over this and the integrity of the structure starts to move. On top of your two or three cheeses place a final pressing rack and on top of this place one or two large solid blocks of timber. This will take the pressure of the car jack on top.

The car jack needs to fit on top of this pressing block and underneath the top beam of wood at the top of the two uprights. As you open the jack up, and it begins to meet the top restraining beam, the pressure forces the pressing blocks downwards onto the cheeses and forces the juice to run from them into the tray and down into the bucket.

You must wait for the juice to stop flowing before increasing the pressure on the jack. When the juice ceases to flow you can reverse the jack, remove everything. The spent pomace can be composted or fed to livestock and the press cloths washed and put away. Wash the former, the tray and the racks, allow to dry and then store. Then gently rinse the wooden structure, and allow to dry and then store.

An every-day food processor is good to use as a small scale scratter or pulper at the start, then you can upgrade to making or buying something bigger.

If you'd like to pasteurise, you can use a heavy pan with a trivet at the bottom and a thermometer inside one uncapped bottle, or you can buy a small pasteuriser from a specialist supplier. If you want to start pasteurising on a large scale, then you can contact me through social media and ask for some photos and help re making a large bath pasteuriser.

Converting Measures

The measures below are British imperial converted to metric. Remember a lot of recipes use American measures, which are different from what are used in the UK.

1 pint (UK) = 0.56 litres

1 quart (UK) = 1.13 litres

1 gallon (UK) = 4.54 litres

1 pound (UK) = 0.45 kilograms

Further Information

Apple tree supplier - Adams Apples -
https://www.adamsappletrees.co.uk/

Apple tree supplier - Keepers Nursery -
https://www.keepers-nursery.co.uk/

Apple tree supplier - Frank Matthews Trees -
https://www.frankpmatthews.com/

Apple rootstock suppliers - widely available on the internet.

Bag in Box suppliers - Jigsaw
http://www.baginboxonline.co.uk/

Bag in Box supplier - Bag in Box shop UK
https://www.baginboxshop.co.uk/

Bottle Company (South) Ltd
www.bottlecompanysouth.co.uk

Cider Workshop This fantastic online resource can be found at
www.ciderworkshop.com

Courses - Cider making courses - many around the country, but
to learn the best try Peter Mitchell's Cider and Perry Academy -
https://www.cider-academy.co.uk/peter-mitchell/   He
even does one day courses now. I can recommend him – I've
been on two of his extensive courses.

Courses – beer and cider -
https://www.beerandcideracademy.org/cider-courses

Courses – wine and beer - http://www.brew-school.com/courses/wine-making-and-viticulture-course-(weekend)/

Courses - Wine courses - https://www.plumpton.ac.uk/courses/wine-division/wine-part-time-courses/

Courses - Wine making courses - https://mantelfarmshop.co.uk/Training-Courses-and-Events/Rural-and-Craft-Courses/Country-Wine-Making-Course

Homebrew Centre - homebrew suppliers for beer, cider & winemakers https://www.homebrewcentre.co.uk/

Homebrew Forum UK - https://www.thehomebrewforum.co.uk/

Homebrew suppliers - Kit & equipment suppliers for beer, cider & wine making https://www.brewuk.co.uk/cider-kits.html

Homebrew Talk - an online forum for beer, wine, cider and mead makers https://www.homebrewtalk.com/

Information about varieties of apples www.orangepippintrees.co.uk

Information about cider apples in France http://www.pommiers.com/pomme/pommier-cidre.htm

Lea, Andrew Craft Cider Making (2nd edition) The Good Life Press Limited(2010)

Orchard Network UK - a directory of orchard groups, stockists etc.
https://www.orchardnetwork.org.uk/regional-directories

The Orchard Project - set up to help groups set up community orchards, and train people in orchard management - **www.theorchardproject.org.uk**

Umpelby, Roger & Copas, Liz. Growing Cider Apples: A Guide to Good Orchard Practice. NACM (2002)

Vigo Ltd This is the UK leading cider equipment suppliers based in Honiton, Devon. **www.vigopresses.co.uk**

GLOSSARY

ABV – Alcohol by Volume. A measurement of the alcoholic content of a beverage expressed as a percentage; such as e.g., 6% or 6%ABV, or ABV 6.

Acid blend – 50% malic acid, 40% citric acid and 10% tartartic acid.

Ascorbic Acid – Also known as Vitamin C. Used to prevent apples turning brown when making juice (NOT cider).

Back Sweetening – Sweetening cider after fermentation has ended by adding apple juice or artificial sweeteners.

Bag in Box, or BIB – A method of storage and dispensing cider or wine, consisting of an inner bag housed in a rigid cardboard box.

BIAB – (Brew in a bag) Whether you're looking for a shorter, simpler all-grain brew day or an extract brewer seeking marked improvement in the quality of your beer without complicating your process, BIAB saves time and money by providing a low cost all grain brewing option.

Bottle conditioning – A naturally sparkling drink (usually cider, beer or wine), occurring when the bottle has been primed with a small amount of sugar prior to filling with a partially fermented alcohol.

Campden tablets – Widely available tablets containing a measurement of sodium metabisulphite sufficient to make

50ppm (parts per million) of sulphur dioxide when dissolved in 1 gallon (4.5 Litres) of liquid. If you are allergic to sulphites then do not use Campden tablets. If a recipe requires you to stop the fermentation by killing the yeast try pasteurising. You can't pasteurise a sparkling drink like beer at home.

Carbonation – the artificial injection of carbon dioxide into a liquid under pressure.

Cheese – a layer of apple pulp, usually encased in a porous cloth and separated from the next cheese by a pressing board or rack.

Ciderkin – A weak cider, usually less than 4.5 ABV, usually made by wetting spent pomace with water and repressing.

Citric Acid – A naturally occurring fruit acid found in citrus fruits. If you need to add citric acid simply use lemon juice.

Cream of tartar – the old name for potassium bitartrate. Can be substituted with Baking Powder (1.5 tsps of baking powder instead of 1 tsp of Cream of tartar).

Cultivar – All trees vegetatively propagated from a particular tree; a clone; a particular variety.

Cyser – Apple juice that has honey added to it prior to fermentation.

Demijohn – A large fermentation jar in either glass or food grade plastic, with a narrow top aperture for insertion of a bung and airlock. Normally 1 gallon size (4.5 Litres).

Dry – An alcoholic drink in which all the original sugar has been turned to alcohol.

Fermentation – A biological process where yeast turns sugar into alcohol.

Finings – a product used to clear the finished wine, beer or cider. Isinglass, bentonite, Irish moss (carragheen) all help to flocculate the yeast (make it stick together) and fall out of suspension.

Gallon – old imperial liquid measurement equivalent to 4.5 Litres.

Grafting – the mechanical process of permanent attachment of a scion to a rootstock.

Hydrometer – An instrument to measure the Specific Gravity (S.G. of a liquid).

Keeving – natural and traditional cider making process used mainly in Western UK and Northern France to make a naturally sweet cider.

Maturation – The period after the end of fermentation when subtle biological changes occur to reduce harshness over time in the finished product.

Melomel – Mead flavoured with fruit.

Metheglyn – Mead flavoured with spices or herbs.

Mill – the piece of equipment or process of reducing whole apples into pulp ready for pressing. Also called pulping or scratting.

Pasteurisation – The process of heat-treating to destroy micro-organisms in a liquid to prevent spoilage.

Pectin – Pectin is a complex starch found in many fruits and vegetables that can linger in wines and cider as a haze. Adding pectolase will help breakdown the fruit or vegetable fibre to release more flavour, and it also clears wines after the end of fermentation.

Perry – An alcoholic drink made from fermenting pear juice. Note – pear cider is apple cider with some pear juice or flavouring added after fermentation.

pH – a scale used to measure the degree of acidity and alkalinity of a solution, with 1 being the most acidic and 14 being the most alkaline.

Pomace – the pulp of apples when milled/scratted or pulped. After pressing the spent pomace can be used as a feed additive for poultry, pigs, sheep and cattle.

Precipitated Chalk – Also known as Calcium Carbonate. Available to buy in 50 or 100 g packs from home-brew suppliers.

Racking – the process of removing the finished cider from the lees in a fermentation vessel into a clean vessel using a siphon or pump.

Rootstock – The root system and stem of a tree onto which a scion or variety is grafted.

Scion – A shoot or twig of a specific variety which is grafted onto the rootstock.

Scratter – A type of mill which breaks down larger fruits such as apples or pears into smaller pieces prior to pressing.

Scrumpy – Old fashioned name given to traditionally made ciders usually from the West Country (from Somerset down to Cornwall).

Seedling – a tree that has grown on its own roots from a seed, rather than a grafted tree with a known variety.

Sodium metabisulphite – A chemical containing sulphur dioxide used to sterilise equipment used for home brewing and in cider making.

Sparging – is the rinsing of the mash grains to extract as much of the sugars from the grain as possible without extracting puckering tannins from the process. Typically, 1.5 times as much water is used for sparging as for mashing.

Specific Gravity or S.G. – A measure used to determine the sugar content of a liquid by use of a hydrometer to estimate the potential final alcohol.

Tannin – A complex chemical compound found in many fruits which impart a range of flavours and mouth-feel to a finished beverage (e.g. cider or sloe gin). Tannins have a bitter taste and an astringent mouth-feel. If you need tannin in a recipe simple make a cup of strong black tea.

Verjuice – A product of non-fermented unripe apples or grapes and used as a condiment.

Vitamin C – Ascorbic Acid. Used to stop apples oxidising when pressing to make juice (not when making cider).

Yeast – Micro-organisms responsible for fermentation. There are many commercial varieties available suitable for cider, wine and beer making, as well as wild yeasts present in the air.

INDEX

213

Printed in Dunstable, United Kingdom

63751752R00121